Cosmic Catastrophes

Supernovae, Gamma-Ray Bursts, and Adventures in Hyperspace

J. Craig Wheeler
University of Texas

CAMBRIDGE
UNIVERSITY PRESS

PUBLISHED BY THE PRESS SYNDICATE OF THE UNIVERSITY OF CAMBRIDGE
The Pitt Building, Trumpington Street, Cambridge, United Kingdom

CAMBRIDGE UNIVERSITY PRESS
The Edinburgh Building, Cambridge CB2 2RU, UK   http://www.cup.cam.ac.uk
40 West 20th Street, New York, NY 10011-4211, USA   http://www.cup.org
10 Stamford Road, Oakleigh, Melbourne 3166, Australia
Ruiz de Alarcón 13, 28014 Madrid, Spain

First published 2000

Printed in the United States of America

Typefaces Sabon 10/13 pt. and Antique Olive   *System* DeskTopPro$_{/\text{UX}}$ [BV]

A catalog record for this book is available from the British Library.

Library of Congress Cataloging in Publication Data
Wheeler, J. Craig.
    Cosmic catastrophes : supernovae, gamma-ray bursts, and adventures in
hyperspace/ J. Craig Wheeler.
      p.   cm.
    Includes index.
    ISBN 0-521-65195-6 (hb)
    1. Supernovae.   2. Stars.   3. Hyperspace.   I. Title.

QB843.S95   W48 2000
523.8'4465–dc21
                                                                99-053321
    ISBN 0 521 65195 6   hardback

Cosmic Catastrophes

Supernovae, Gamma-Ray Bursts, and Adventures in Hyperspace

In this tour de force of the ultimate and extreme in astrophysics, renowned astrophysicist and author J. Craig Wheeler takes us on a breathtaking journey of supernovae, black holes, gamma-ray bursts and hyperspace. This is not far-fetched science fiction but an enthusiastic exploration of ideas at the cutting edge of current astrophysics.

When a massive star has exhausted its nuclear fuel, it comes to the end of its life. Rather than dying quietly, with a whimper, it erupts in a catastrophic explosion known as a supernova. So much energy is released that a single supernova can outshine an entire galaxy of billions of stars. Taking these spectacular events as a linking theme, this book follows the tortuous life of a star – from birth to death – and leads on to ideas of complete collapse to a black hole, worm-hole time machines, the possible birth of baby bubble universes, and the prospect of a revolution in our view of space and time with a ten-dimensional string theory. Along the way, we look at evidence that suggests that the Universe is accelerating, and the recent revolution in understanding gamma-ray bursts – perhaps the most catastrophic cosmic events of all.

With the use of lucid analogies, simple language, and the crystal-clear illustrations of Tim Jones, this book makes accessible some of the most exciting and mind-bending objects and current ideas in the Universe.

J. Craig Wheeler is the Samuel T. and Fern Yanagisawa Regents Professor of Astronomy at the University of Texas at Austin, where he was chair of the department from 1986 to 1990. In 1999, he received both the Dad's Association Teaching Fellowship and the President's Associates Teaching Excellence Award from the University of Texas and was elected as Vice President of the American Astronomical Society. He has edited books on supernovae and accretion disks and published a novel, *The Krone Experiment*. For recreation, he enjoys his old college sport of fencing, running, reading, and writing fiction.

To my sons,
Diek W., the scientist,
and J. Robinson, the artist.

Contents

Preface

The core of this book concerns supernovae, my principle research interest, but the broader theme is the connection of these cosmic catastrophes with the sweep of intellectual ferment in astrophysics. The story leads from the birth, evolution, and death of stars to the notion of complete collapse in a black hole, to worm-hole time machines, the possible birth of new universes, and the prospect of a conceptual revolution in our views of space and time in a ten-dimensional string theory. It is all one glorious, interconnected Universe, both physically and intellectually. Or maybe there are more than one.

In terms of astrophysical connections, the book reaches back to the origins of stars and how they evolve, treats the mechanisms of supernovae, and then moves forward to the compact progeny of supernovae – neutron stars and black holes. Neutron stars are presented in all the variety we know today – pulsars, millisecond pulsars, binary pulsars, magnetars, and X-ray sources both steady and transient. The concrete manifestation of black holes in observational astronomy, especially in binary stellar systems, is described. Topics that have come to light as the book was being written, soft gamma-ray repeaters and the revolution in cosmic gamma-ray bursts, are presented. The scientific background is given to understand what kind of supernovae are used to produce the radical notion of the acceleration of the

Universe, and how and why. Similar background aids in making the connection between flaring gamma-ray sources and compact objects.

A parallel theme is not the objects themselves, but the intellectual framework that underlies our study and the limits to which it currently extrapolates. This involves discussions of the physics of the twentieth century, the quantum theory and Einstein's gravity, how they collide, and the prospects for reconciliation. In the process, the concept of gravity as curved space is shown to lead to radical notions such as time machines and baby bubble universes. The promise of string theory to give a unifying view and to open new conceptual windows is illustrated.

Because I have used and intend to use this book for classes, I have, for completeness, written about topics that have been presented before: the basics of stellar evolution, the discovery and interpretation of pulsars, the nature of space and time in the vicinity of black holes, and the more recent topics such as worm holes and the promise of string theory. I have presented this material in my own style and hope that there is some benefit to seeing it again. In addition, I have tried to present this material in a broad context that gives it a different perspective than previous treatments.

There are other topics that I have stressed here because they are of crucial importance and because they tend to get overlooked. One of these is binary star evolution. When I began to teach this material, there was scarcely any mention of binary stars in introductory astronomy texts, save perhaps for a mention of eclipses and visual and spectroscopic binaries. Current texts are much better, but this topic is so fundamental that I am compelled to present it in some detail. Supernova researchers believe many supernovae depend incidentally or critically upon their being in binary systems. Much of what we know about neutron stars follows from their being in binaries. The only way we know about stellar mass black holes is by discovering them in binary systems. Many books on black holes concentrate on the supermassive variety in galactic nuclei and scarcely mention those in binary systems, never mind the amazing array of phenomenology associated with them and the reasons for it. I have thus devoted a chapter to discussing the systematics of Roche lobes, mass transfer, and common envelopes, the language of this field that is often passed over in books of this kind.

A closely related topic is that of accretion disks. The study of disks has become an industry unto itself, but these objects are rarely presented with the background of how they work and why they are so important to the topics of this book, from the evolution of Type Ia supernovae to binary neutron stars to binary black holes to the cosmic gamma-ray bursts. Accretion disks have a life of their own with instabilities that cause them to flare and attract the attention of astronomers. With the exception of venerable old Cygnus X-1 and a few others, all the host of new black-hole candidate discoveries are due to flaring systems. The most plausible mechanism for the flaring is associated with the disk. Accretion disks also merit a separate chapter.

I have also included topics that, although the subject of many articles in the popular science literature, have not, to my knowledge, been incorporated in a book where the relevant background can be laid out in advance and the story told as an integral part of modern astrophysics. There are three examples of that, all of which have "exploded" in the past year. One is the proof that the soft gamma-ray repeaters involve exceedingly strongly magnetized neutron stars – magnetars in the language of my colleague Robert Duncan. Another story is the amazing array of developments that have followed since the discovery of the first optical counterparts of the cosmic gamma-ray bursts, not the least of which, to someone of my bent, is the association of one with a supernova. In each of these cases, to understand the story behind the headlines fully, one needs to know the relation of the topic to stellar evolution, the ideas behind the birth of neutron stars and black holes, the significance of supernovae that show a paucity of hydrogen and helium, and the nature of binary star evolution. Last, but certainly not least, is the use of supernovae to measure distances on cosmological scales. The tentative result, that the Universe is accelerating, was recently proclaimed the scientific breakthrough of the year 1998 by *Science Magazine*. Here I have the opportunity to tell the story in terms of the history of the topic as well as the astrophysical background involving binary star evolution, specific supernova mechanisms, and the elements of cosmology.

The seeds of this book were planted in 1975. My colleague, R. Edward Nather, invented a course at the University of Texas called Astronomy Bizarre. The purpose of this course was to tell the story of the Universe from the big bang onward, rather than from the Solar System outward as is traditional for introductory astronomy courses, and to introduce some of the exotica of astronomy for which one has little time in the standard introductory course for nonscience majors.

Nather taught the first version of this course just after I arrived at the University of Texas. The prerequisite of a standard introductory astronomy course was omitted from the catalog. More than 300 students registered, and a second section had to be opened. I was assigned that section and have been teaching some version of the course for the last 25 years. This book represents some of the material I have developed for the course.

Nather and I planned to write a book based on his original Astronomy Bizarre syllabus. We wrote a draft, but the project foundered for various reasons. The material that ended up in this book is very different from that first draft, but the early introduction of the notion of conserved quantities is a vestige of that work, and I thank Ed for that idea.

Astronomy Bizarre was such a successful course that it evolved to encompass several versions. Over the years, I inherited the course that concentrated on stars. To keep my teaching fresh, I have regularly changed the content of the course. Sometimes I concentrate on supernovae and closely related topics. Other times, I have taught the whole course just on black holes and related ideas. I have taught it sometimes to a small class required to do substantial writing. To stay current, I have added new material as new developments have come along, a never-ending process in astrophysics.

As I have taught the course, I have had to wrestle with how to portray the complex and fascinating ideas of astrophysics to classes of bright, interested, but nontechnically trained students. This book also represents a compilation of the ideas I like to try to explain to popular audiences and the techniques I have developed to accomplish this. One of the ideas with which I am most pleased is blowing up a balloon and turning it inside out to portray the embedding diagram of the curved space around a black hole. I have also tinkered with the vocabularly. In many cases, I adopt the jargon of astronomy and endeavor to define and explain it. In other cases, I have invented new phrases. I did not think that the term "degeneracy" carried much import for a popular audience, even after an attempt to explain it. I have thus referred to a "quantum pressure" rather than "degeneracy pressure," feeling that this term gets the basic point across that this pressure is different in a fundamental way from that exerted by a gas of hot plasma. I trust that these attempts to make the material accessible to non-science-major students have some value for audiences beyond the lecture hall.

In addition to the various themes of the book I outlined earlier, I have emphasized several physical themes that tie together various topics of the course. I stress the difference between stars supported by thermal pressure and those supported by the quantum pressure, why one results in regulated nuclear burning and one leads to stellar explosions. These lessons are used throughout stellar evolution from star formation to hydrogen burning to red-giant formation to the formation of iron cores and the contrasting examples of classical novae and Type Ia supernovae. The nature of the weak interaction and the intimate connection to neutrinos is introduced early and used to relate the topics of the solar neutrino problem, massive core collapse, and the radioactive decay that powers the light curves of supernovae devoid of extended envelopes of matter at the time of explosion.

Over the years, many friends and colleagues have helped me to understand the material I have tried to synthesize in this book. Any errors of fact or interpretation are mine, not theirs. I am indebted to Ed Fenimore for clarifying the early history of gamma-ray bursts. Special thanks go to Stirling Colgate for his contributions to the research depicted here and for his intensity and wide-ranging imagination that have stimulated me both scientifically and otherwise.

I am grateful to all my students over the years as I have developed and altered the course. Their feedback has allowed me to better understand what works and what does not. In the spring of 1998, I made this feedback more concrete by offering extra credit to students in my Astronomy Bizarre class who would make comments on clarity and errors in the draft of the book I was using for class. Many of them made very valuable suggestions that I have incorporated. Among these people were Ramesh Dhanaraj, Angela Entzminger, Laura Tamayo Gamborino, John Going, Jonathan Hurley, John Kendall, Sara Keyes, Rubi Melchor, Siddarth Ranganathan, Natalie Sidarous, Benjamin Tong, and Victor Yiu.

I am also grateful to Adam Black of Cambridge University Press for his enthusiasm for this book and especially to Timothy Jones whose magic with computer illustration has brought many ideas to life.

1

Setting the Stage

Star Formation and Hydrogen Burning in Single Stars

1. Introduction

We look up on a dark night and wonder at the stars in their brilliant isolation. The stars are not, however, truly isolated. They are one remarkable phase in a web of interconnections that unite them with the Universe and with us as human beings. These connections range from physics on the tiniest microscopic scale to the grandest reaches in the Universe. Stars can live for times that span the age of the Universe, but they can also undergo dramatic changes on human time scales. They are born from great clouds of gas and return matter to those clouds, seeding new stars. They produce the heavy elements necessary to make not only planets but also life as we know it. The elements forged in stars compose humans who wonder at the nature of it all. Our origin and fate are bound to that of the stars. To study and understand the stars in all their manifestations from our life-giving Sun to black holes is to deepen our understanding of the role of humans in the unfolding drama of nature.

This book will focus on the exotica of stars, their catastrophic deaths, and their transfigurations into bizarre objects like white dwarfs, neutron stars, and black holes. This will lead us from the stellar mundane to the frontiers of physics. We will see how stars work, how astronomers have come to understand them, how new

1

knowledge of them is sought, how they are used to explore the Universe, and how they lead us to contemplate some of the grandest questions ever posed.

We will begin by laying out some of the fundamental principles by which stars and, indeed, the Universe function.

2. Background

2.1. The Basic Forces of Nature

The nature of stars is governed by the push and pull of various forces. The traditional list of the basic forces of nature is as follows:

- *Electromagnetic Force* – long-range force that affects particles of positive (+) and negative (−) electrical charge, as shown in Figure 1.1 (top). *Protons* (p) are examples of positive charges, and *electrons* (e−), negative charges.
- *Strong or Nuclear Force* – short-range force that affects heavy (high-mass) particles such as protons (p) and *neutrons* (n). The strong force binds protons and neutrons together in the atomic nucleus, as shown in Figure 1.1 (middle). The strong force turns repulsive at very small distances between the particles.
- *Weak Force* – short-range force that affects interactions between light (low-mass) particles such as electrons (e−) and *neutrinos* (v). The weak force converts one light particle into another and one heavy particle into another; for instance,

as shown in Figure 1.1 (bottom).
- *Gravity* – long-range force that affects all matter and is only attractive.

The particle known as the neutrino is a special one with no electrical charge. It interacts only by means of the weak force (and gravity),

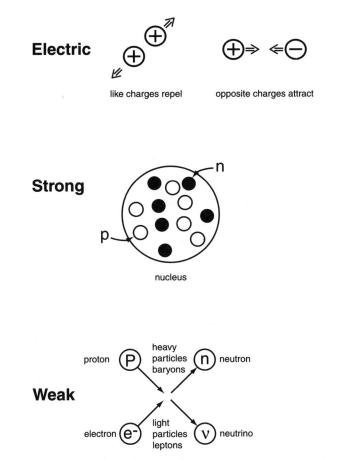

Figure 1.1. The action of the basic forces: (top) opposite electrical forces attract, and like charges repel; (middle) the attractive nature of the strong force holds protons and neutrons together in atomic nuclei despite the charge repulsion among the protons; (bottom) the weak force causes protons to convert into neutrons and electrons into neutrinos and vice versa.

that is to say scarcely at all. Its properties and its role in nature will be explained in more detail below and in later chapters.

The results of theoretical work in the 1960s by Steven Weinberg, Abdus Salaam, and Sheldon Glashow followed by experimental verification in the 1970s and 1980s by a large team led by Carlo Rubbia and Simon Van Der Meer showed that the electromagnetic and weak forces are actually manifestations of the same basic force, which has come to be called the *electroweak force*. This unification is analogous to the recognition based on the work of Thompson and Maxwell in the nineteenth century that electrical effects and magnetic effects are actually intimately interwoven in what we now call the electromagnetic force. Nobel Prizes are only the celebrated tip of the ferment that leads to scientific progress; however, their winners deserve their credit, and the prizes are signposts of major progress. Weinberg, Salaam, and Glashow won the 1979 Nobel Prize in Physics for their work; Rubbia and Van Der Meer, for theirs in 1984.

Current research is aimed at the goal of showing that the strong force is also related to the electroweak force, and that both are manifestations of some yet more fundamental force. Definite progress has already been made toward this goal of constructing a *grand unified theory*. Another dream is to show how gravity may also be understood as intrinsically related to the other forces. The story of gravity is a complex one at the heart of modern physics, and even its role in the pantheon of forces requires some interpretation. Newton interpreted gravity as a force, but, as will be elaborated in Chapter 9, Einstein's theory leads to the interpretation that gravity is a property of curved space and time, that there is no "force of gravity" in the sense that Newton conceived it. Recent dramatic progress has been made toward a unified picture of gravity and the other forces by envisaging particles as one-dimensional strings, rather than as points, as we will see in Chapter 12. In this evolving theory, gravity is again interpreted as a force, but one Newton would scarcely recognize. In practice, we will often refer to these forces in their four traditional categories, as given earlier, with emphasis where appropriate on the interpretation of gravity as a property of curved space.

2.2. Conservation Laws

To a physicist, conservation does not mean careful use to ensure future supplies, but that some quantity is constant and does not change during an interaction. Physicists have learned to make powerful use of principles of conservation, which are stated in roughly the following manner: "I don't care what goes on in detail; when all is said and done, quantity X is going to be the same." Conservation laws do not help to untangle the details of a given physical process; rather, they help to avoid complex details. Conservation laws are of great help exactly when the details are complicated because one can proceed with confidence that certain basic quantities are known and unchanging, despite the details. How this works will be more clear when we see how these conservation laws are used in various ways. They are employed to help understand why stars get hotter when energy is radiated away, the nature of nuclear reactions that power the stars, why stars become red giants and white dwarfs, the very existence and role of the elusive neutrino, how stars circle one another in binary orbits, why disks of matter form around black holes, and why some supernovae shine by radioactive decay. For now we will describe some of the conservation laws most frequently used in the astrophysics of stars.

One of the most fundamental conservation laws is the *conservation of energy*. Energy can be converted from one form to another so understanding energy conservation can sometimes be tricky, but, for all physical interactions, energy is conserved. The energy can be converted from energy of directed motion to random thermal energy and from, or to, gravitational energy. Even mass can be converted to energy and energy to mass according to Einstein's most famous formula, $E = mc^2$. Despite all these potential conversions in form, the energy of a physical system is conserved. When you drop a piece of chalk, it shatters with a small crash, as illustrated in Figure 1.2 (top). The potential gravitational energy goes first into the kinetic energy of falling, then into the energy of breaking electrical bonds among the particles of chalk, and even into the energy of the sound waves of the noise that is made. Despite the complicated details, the total energy of everything is conserved.

Momentum is a measure of the tendency of an object to move in a straight line. The measure of the momentum is not which team scored the last touchdown or goal, a common usage of the phrase in a sports context, but the product of the mass of an object with its velocity. The *mass* is a measure of the total amount of stuff in an object. The *velocity* is the speed in a given direction. Momentum characterized as mass times velocity is also conserved. A mass moving with a certain speed in a certain direction will continue to do so unless acted upon by a force. A given mass may be sped up or slowed down by the action of a force, but the agent supplying the force must suffer an equal and opposite reaction so as to conserve the momentum as a whole. Try jumping suddenly out of a boat (Figure 1.2; middle) and ask your companions if they appreciate the overwhelming verity of the principle of conservation of momentum. If you leap out one side, the boat must react by moving in the opposite direction with the same

Conservation of Energy

Conservation of Momentum

Conservation of Angular Momentum

Figure 1.2. The principles of conservation: (top) dropping and shattering a piece of chalk is a complicated process, but the energy of breaking, motion, heat, and noise is exactly that gained by falling; (middle) a person leaping from a boat will send the boat and his companion rapidly in the opposite direction, illustrating conservation of momentum; (bottom) a skater drawing in his arms will spin faster, conserving angular momentum.

momentum as your leap. The boat will inevitably tip and leave everyone in the drink.

Angular momentum is a property related to ordinary momentum, but it measures the tendency of an object of a given mass to continue to spin at a certain rate. The measure of the angular momentum is the mass times the velocity of spin times the size of the object. A popular demonstration of conservation of angular momentum is an ice skater. When a spinning skater draws his arms in closer to his body, his "size" gets smaller. Because his mass does not change, his rate of spin must increase to ensure that his total angular momentum will be constant. In detail, this is a complex process involving the contraction and torsion of muscles and ligaments. You do not have to understand the details of how muscles and ligaments work, however, to see that the skater must end up in a dizzying spin when he pulls his arms in, and that he will slow again by simply extending his arms (Figure 1.2, bottom).

Other conservation laws are important to physics but are not reflected so easily in everyday life. An especially powerful example is that of *conservation of charge*. Electrical charge, the total number of positively and negatively charged particles, is conserved. Physical processes can cancel charges, a positive charge against a negative one, but the net positive or negative charge cannot change in a physical process. Neither positive nor negative charges can simply appear or disappear. In a reaction involving a bunch of particles, the total charge at the end of the reaction must be the same as at the beginning of the reaction. Here is an example:

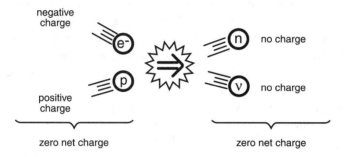

Elementary particles have other properties, akin to electrical charge, that are conserved. The heavy particles like protons and neutrons that constitute atomic nuclei are called *baryons* (from the Greek "bary" meaning heavy). In a nuclear reaction, the number of baryons is conserved. The baryons may be changed from one kind to another, protons to neutrons for instance, but the number of baryons does not change. If there were four baryons at the start, there will be four at the end. The same example applies to baryons:

There are other elementary particles that do not belong to the baryon family. The ones in which we will be especially interested are the low-mass particles known as *leptons*. Electrons and neutrinos are members of this class. As for baryons, nuclear reactions conserve the total number of leptons, even though individual particles may be created or destroyed. Common reactions will involve both baryons and leptons, and both classes of particles are separately conserved. That is true in our sample reaction:

These last two conservation laws, of baryon number and lepton number, are highly accurate. These laws were once thought inviolate. Recent theoretical developments have suggested that this is not strictly true. One of the suggestions arising from the work of constructing a grand unified theory of the strong and electroweak forces is that baryons may not be completely conserved. The big bang itself may depend on the breakdown of these conservation laws. On time scales vastly longer than the age of the Universe, baryons, including all the protons and neutrons that make up the normal matter of stars, may decay into photons and light particles. For all "normal" physics, and hence for all practical purposes, baryons and leptons are conserved, and we will use these conservation laws to understand some of the reactions that are crucial to understand the nature of stars.

An important offshoot of the ideas of conservation of energy, charge, baryon number, and lepton number is the existence of matter and antimatter. For all ordinary particles – electrons, neutrinos, protons, and neutrons – there are antiparticles – antielectrons, antineutrinos, antiprotons, and antineutrons. These are not fantasy propositions; they are made routinely in what are loosely called "atom smashers," and more formally, particle accelerators, and they rain down continually on the Earth in the form of cosmic rays. The connection to the conservation of charge is that antiparticles always have the opposite charge of the "normal" particle. The antielectron, also called a *positron*, has a positive electrical charge. An antiproton has a negative charge. Because neutrinos and neutrons have no electrical charge, neither do their antiparticles, but they have other complementary properties. For instance, to make sense of the way physics works, it is necessary to consider an antielectron to count as a "negative" lepton and an antiproton to count as a "negative" baryon. In that sense, assigning the property of "leptonness" or "baryonness" to a particle is like assigning an electrical charge; it can be positive or negative and is opposite for particles and their antiparticles.

A remarkable property of particles and antiparticles is that they can be produced from pure energy and can annihilate to produce pure energy. Carl David Anderson won the Nobel Prize in Physics in 1934 for the discovery of positrons. Positrons were first created in a laboratory by applying a very strong electric field, the energy source, to an empty chamber, a vacuum. When the electric field reached a critical value, out popped electrons and positrons. You can see the connection with conservation of energy, charge, and leptons here. The energy of the electric field must be strong enough to provide the energy equivalent of the mass of an electron and a positron, twice the mass of a single electron. Because the original vacuum, even with the imposed electrical field, had no net electrical charge, the final product, the electrons and positrons, also must have no net electrical charge. For every negatively charged electron that is created in this way, there must be a particle with the opposite electrical charge, an antielectron, a positron. Likewise, the original apparatus had no "leptons," just the electrical field and vacuum. When an electron and positron appear, the electron must count as plus one lepton, and the positron as minus one lepton, so that the net number of leptons is still zero, in analogy with the way one keeps track of electrical charge. Here is a schematic reaction:

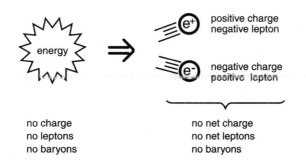

This experiment can also be run backward. If an electron and positron collide, they annihilate to produce pure energy – photons of electromagnetic energy – with no net electrical charge and no net number of leptons. The same is true of any particle and antiparticle. When they collide, they annihilate and produce pure energy; all the mass disappears. This is a very dramatic example of conservation of energy and of Einstein's formula, $E = mc^2$; pure energy can be converted into matter, and matter can be converted into pure energy. In the process, the total number of electrical charges, the total number of leptons, and the total number of baryons does not change. The total of each is always zero.

You might wonder, if antiprotons annihilate protons on contact and hence are antimatter, do they antigravitate? If I make an antiproton in a particle accelerator, will it tend to float upward? The answer is no. Energy is directly related to mass by the formula $E = mc^2$. One implication of this relation is that because mass falls in a gravitational field, energy also falls in a gravitational field. Because particles and antiparticles annihilate to form a finite, positive amount of energy that will fall in a gravitational field, so the individual particles and antiparticles must fall. An antigravitating particle might annihilate with a gravitating particle to produce no energy, but we do not know of any such particles. Current physics does give some hints of the existence of antigravity which we will discuss in Chapter 12.

2.3. The Energy of Stellar Contraction

We can now apply these various conservation laws to stars. We will start with the principle of conservation of energy. The result is a little surprising at first glance, but crucial to understanding the way in which stars evolve.

Let us first consider the nature of a star. A star is a hot ball of gas in *dynamic equilibrium*. This means that a pressure of some kind pushes outward and balances the gravity that pulls inward. The Sun does not have the same size day after day because there are no forces on it that might alter its size; rather there are great forces both inward and outward at every point in the Sun. The structure of the Sun has adapted so that the forces just balance. The equilibrium is such that the pressure force keeps gravity from collapsing the star, and gravity keeps the pressure from exploding the star. We will see in Chapter 6 that this condition of delicate balance can be interrupted and either collapse or explosion can result, depending on the circumstances. The mass and size of a star determine the gravity and hence the pressure and heat needed to arrange the balance of forces.

The Sun and most stars we see scattered in the night sky are supported by the pressure of a hot gas. The pressure, in turn, is directly related to the thermal energy in the star. At the same time, the star is held together by gravity. As the star radiates energy into space, it loses a net amount of energy. What happens to the temperature in the star? The answer is dictated by the principle of conservation of energy.

If the star were like a brick, the answer would be simple. As energy is radiated away, a brick just cools off. Gravity plays a crucial role in the makeup of a star, however. If the star were to cool, the pressure would tend to drop, and then gravity would squeeze the star, compressing and heating it. A star responds to a loss of radiant energy in just this paradoxical way. As the star loses energy, it contracts under the compression of gravity and actually heats up! This process, illustrated in Figure 1.3, is completely in accord with the conservation of energy. One must remember only that the squeezing by gravity is an important energy source that cannot be ignored when counting up all the energy, just as the energy of falling breaks the chalk in Figure 1.2.

If nuclear reactions happen by accident to momentarily put more energy into the star than it radiates, the star gains energy. What happens to its temperature in this case? If you were bitten in the first case, you should be wise by now. As shown in Figure 1.3, if the temperature were to go up, the pressure would rise and push outward against gravity. The expansion would cause the star to cool. That is just what a star does; if you put in an excess of energy, it expands and gets cooler.

This apparently contradictory behavior of a star to heat up when it loses energy and cool off when it gains energy is a direct application of the law of conservation of energy. This behavior is crucial for the evolution of stars as various nuclear fuels flare up and burn out.

2.4. Quantum Theory

Things work differently in the microscopic world of atoms and elementary particles than would seem to be "normal" from our everyday experience. On the scale of very small things, behavior is described by *quantum theory*. On this scale, changes do not occur smoothly, but in jumps. The behavior of matter on the quantum level does, however, have important implications for big things like stars.

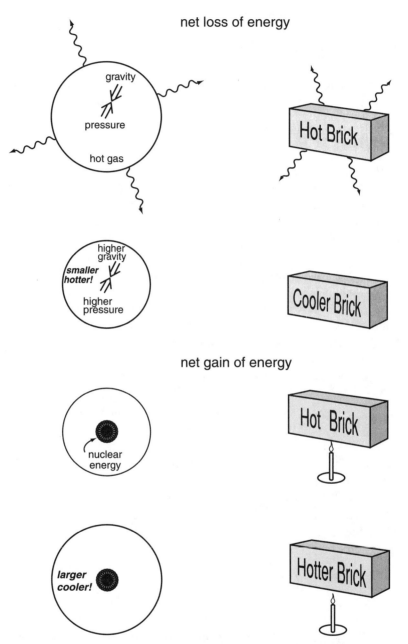

Figure 1.3. Stars supported by the pressure of a hot gas behave differently than a solid object like a brick. A brick will cool off as it radiates energy and heat up if a source of energy is added. Because of the action of gravity, a star held up by the thermal pressure of a hot gas will heat up when it loses a net amount of energy by radiation and cool off if it gains a net amount of energy from nuclear reactions.

In our ordinary macroscopic world, the old argument about the impenetrability of matter is approximately true; you cannot put your fist through a concrete wall. Your fist and the wall cannot occupy the same volume. The notion of impenetrability is very different in the microscopic world of the quantum theory. According to the quantum theory, elementary particles are not hard little balls, but also have wavelike qualities to them. Particles can, in principle, occupy exactly the same position in the same way that two ripples on a pond can occupy the same position momentarily as they pass through one another. Another aspect of the wavelike nature of particles is that their position cannot be specified. Think of the task of specifying where an ocean wave is: is it where the surface starts to curl upward, where the froth breaks on the crest, or in the wake? According to the *uncertainty principle* of the quantum theory, the positions of particles cannot be specified exactly. More precisely, there is complementary uncertainty between the position and the momentum of a particle. If the location of a particle is limited in some way, for instance by being confined in an atom, the momentum and the energy become very uncertain. If the momentum is made more certain, you do not know where the particle is. According to the quantum theory, the position of a particle is the place where it might be and the volume it occupies is a measure of the uncertainty of its position. Rather than hard spheres, particles are more like little fuzzy balls or collections of waves. This property of uncertain position, momentum, and energy allows more than one of them to occupy the same volume in the right circumstances.

There are particles in the quantum world, however, that in special situations possess a property of absolute impenetrability. Among the particles that possess this property are familiar ones – electrons, protons, neutrons, and neutrinos. Particles of this class cannot occupy the same little smeared-out uncertain region of space if they have the same momentum, or, rather loosely, the same energy. This property is known formally in the quantum theory as the *exclusion principle*. Curiously, these particles can occupy the exact same volume as long as they have different momentum or energy. Two electrons, for instance, cannot occupy the same place if they have the same momentum, but they can if they have different momentum, as shown in Figure 1.4 (top). A common particle that does not obey the exclusion principle is the photon; two photons of the same energy can occupy the same volume at the same instant.

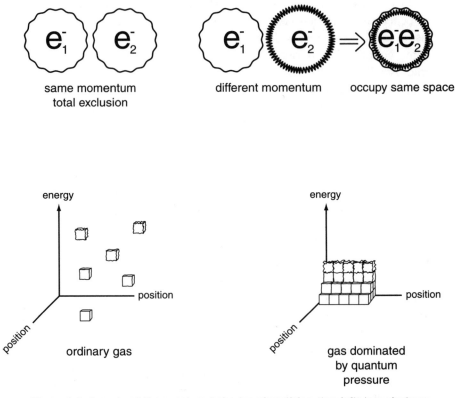

Figure 1.4. Aspects of the quantum behavior of particles: (top left) two electrons with the same momentum are absolutely excluded from being in the same place, from occupying the same volume; (top right) if one electron has a different momentum and hence energy, its "waves" are in a different state and this allows the two electrons to occupy exactly the same volume; (bottom left) a normal gas of hot particles has the particles spread out in position and energy so that quantum effects are not important and the resulting thermal pressure depends on the temperature as well as the density of the particles; (bottom right) if particles are packed tightly enough by having a very high density, then particles with the same energy occupy volumes dictated by the uncertainty principle but, according to the exclusion principle, cannot occupy the same volume unless they have different energies. The energy acquired by the particles depends only on the density and not the temperature, but it can provide a quantum pressure that can support a star.

The uncertainty and exclusion principles determine the structure of atoms. The electrons exist in a smeared volume surrounding the atomic nucleus. The size of this volume is in accord with the uncertainty principle and the fact that electrons are wavelike and their positions cannot be specified precisely. The electrons are confined into a restricted volume by the positive attraction of the protons in the nucleus. The electrons can all occupy nearly the same volume because some have higher energy, thus satisfying the constraint of the exclusion principle.

These quantum properties of particles come into play in a very important way as stars evolve. Normally the particles in a star are spread out in space and in energy, as shown in Figure 1.4 (bottom left). In this situation, the gas exerts a *thermal pressure* as the particles randomly collide and bounce off one another and generally tend to move apart. This thermal pressure associated with a hot gas supports the Sun and stars like it.

As the stars burn out their nuclear fuels, they contract and become very dense. The electrons in the stars are squeezed tightly together. The electrons get compacted into a state where the volume of quantum uncertainty occupied by each electron is bumping up against that of its neighbor. Electrons of the same energy would then absolutely resist any more compaction. That state of the star would be the maximum compression allowed according to the exclusion principle if no two electrons could occupy the same volume. Many electrons can, however, occupy the same volume if some of the electrons have extra energy. Extra energy does arise in this circumstance as a result of the compaction by gravity and the action of the uncertainty principle. As the space that the electrons occupy becomes more confined, their positions becomes more "certain." To satisfy the uncertainty principle, the energy (strictly speaking the momentum) must become more uncertain. As the uncertainty in the electron energy becomes higher, the effective average energy of the electron increases. Thus the compaction squeezes the electrons together, the exclusion principle prevents two electrons with the same energy from occupying the same volume, and the restricted volume gives the electrons more energy in accord with the uncertainty principle. With more energy, some electrons can now occupy the same volume as illustrated in Figure 1.4 (bottom right). The fact that electrons can gain energy and hence overlap in the same volume allows greater compaction of the star.

The net effect is that the squeezing of the electrons gives them an energy that derives purely from quantum effects. The "quantum energy" that results from stellar compaction depends only on the density and is completely independent of the temperature. This quantum energy can exceed the normal thermal energy due to random motion of gas particles by great amounts. The electrons that acquire this quantum energy can also exert a *quantum pressure*. This quantum pressure can provide the pressure to hold the star up even when the thermal pressure is insufficient.

The fact that the quantum pressure is independent of the temperature has major implications for the thermal behavior of compact stars for which this pressure dominates. When a star is supported by the quantum pressure, it does not contract upon losing energy by radiating into space. The reason is that as the temperature drops, the quantum pressure is unaffected and remains constant. A star supported by quantum pressure behaves like "normal" matter; when it radiates away energy, it cools off. In this sense, such a star is more like a brick that just cools off when it radiates its heat, as illustrated in Figure 1.3.

Stars supported by the quantum pressure of electrons are known as *white dwarfs*. They will be discussed in more detail in Chapters 5. Only so much mass can be supported by the quantum pressure of electrons. This limiting mass is called the *Chandrasekhar mass* after the Indian physicist, Subramanyan Chandrasekhar, who first worked out the concept, shortly after the birth of the quantum theory. Chandrasekhar did this work as a very young man and was finally awarded the Nobel Prize for it in 1983. Chandrasekhar and his work have been honored once again by naming a major NASA orbiting observatory, the *Advanced X-ray Astronomy Facility*, the *Chandra Observatory*. The maximum mass a white dwarf can have for an ordinary composition is 1.4 solar masses, not much more massive than the Sun. If mass were to be piled onto a white dwarf so that its mass exceeded that limit, the white dwarf would collapse, or perhaps explode if it were composed of the right stuff. That notion will be explored in Chapter 6.

3. Evolution

The mass of a star sets its fate. The structure and evolution of a star of typical composition follow from the mass with which it is born. The mass determines the pressure required to hold the star up. The condition that the pressure balances gravity determines the temperature and the temperature determines the rate of nuclear burning and hence the lifetime of the star. For much of a star's life, the pressure to support it comes from the thermal pressure of a hot gas. This means that when a star loses a net amount of energy it heats up and when it gains a net amount of energy it cools off, as described in Section 2.3. This fundamental property controls the development of the star.

3.1. Birth

Stars first come into existence as protostars. Protostars are thought to form by some sort of intrinsic instability in the cold molecular gas that pervades the interstellar medium. Sufficiently massive clumps of this matter have an inward gravity that exceeds the pressure they can exert, so they contract and become ever more dense and hot until nuclear reactions start and the clump becomes a star. Alternatively, there are processes involving energetic shock waves that may cause the matter floating through space to clump together. The shocks may come from the passage of the interstellar gas through the spiral arm of a galaxy, from the explosion of a supernova, or from the flaring birth of another nearby star.

When a protostar forms, it is not yet hot enough to burn nuclear fuel. To burn nuclear fuel, the protostar must get hotter. The wonderful property of stars, even as protostars, is that if they must become hotter to yield nuclear input, they will automatically do so. That is the nature of the star machine, a machine controlled by conservation of energy under the influence of gravity. For the protostar, this works because the protostar is warmer than the cold space around it. Under this circumstance, the protostar will radiate energy into space. Because a protostar has no energy input from nuclear burning, it loses a net amount of energy into space. This is exactly the circumstance in which a star will heat up! As shown in Figure 1.5, the protostar will continue to lose energy and heat until it becomes hot enough to ignite its nuclear fuel. At this point, the protostar becomes a real star, shining with its own nuclear fire.

3.2. The Main Sequence

If you point at a person in a crowded shopping mall, the probability is that the person is middle aged, neither an infant nor very aged. The stars about us in space have a similar property. If you pick a star in the night sky at random and ask what it is doing, the probability is that it will be in the phase where stars spend most of their active lives. When stars were first categorized, most were empirically found to fall in one category in terms of the basic observable criteria of temperature and luminosity. This category is called the *main sequence*. We

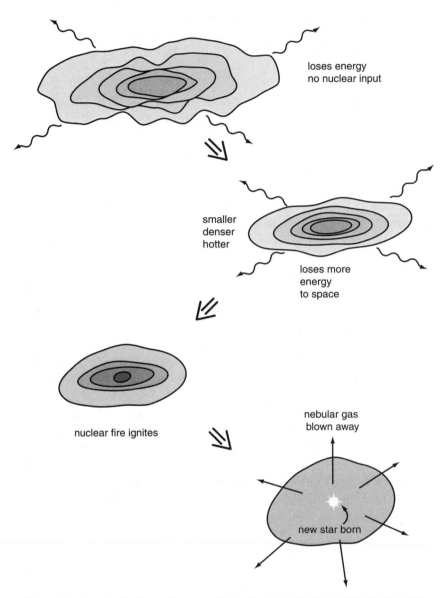

loses energy
no nuclear input

smaller
denser
hotter

loses more
energy
to space

nuclear fire ignites

nebular gas
blown away

new star born

Figure 1.5. A protostar forms from a swirling cloud of cold interstellar gas. Because it radiates into space, but has no nuclear input, the protostar will contract under the pull of gravity and become smaller, denser, and hotter. This process will continue until the center of the star becomes hot enough to light the nuclear fire. The excess gas is blown away and the star emerges from its cocoon to shine with its own nuclear energy.

now understand the physical meaning of the main sequence. Stars are composed mostly of hydrogen, and the main sequence represents the phase of the thermonuclear burning of that hydrogen. Hydrogen burns for a long time compared to other elements. For this reason, stars spend most of their active lifetimes not as protostars or highly evolved stars but as hydrogen-burning stars, just as humans spend most of their lifetime as adults not as infants or octogenarians.

The Sun is in the main sequence hydrogen-burning phase. It is about halfway through its allotted span of 10 billion years. Stars more massive than the Sun burn hydrogen for a shorter time. This may seem strange because massive stars contain more hydrogen fuel to burn. The reason is that massive stars require a greater pressure to support them and hence have a higher temperature. This causes them to burn their extra fuel at a far more prodigious rate than the Sun and so spend their extra fuel in a very short time. Likewise, stars with less mass than the Sun have lower pressure and temperature. They burn their smaller ration of fuel very slowly and live on it far longer than even the Sun will. Stars with less than about 80 percent of the mass of the Sun that were born when the Galaxy first formed have scarcely begun to evolve; the Universe is not old enough yet.

A given star burns its hydrogen at a very steady rate. This is because the star acts to regulate its burning to a very precise level, using the same principle of energy conservation that ignites the fuel in the first place. If the nuclear furnace belches slightly and puts forth a little more heat than can be carried off by the radiation from the star, the excess heat increases the pressure and causes the star to expand. The excess energy is spent in making the star expand. More energy goes into the expansion than was produced in the nuclear belch, and the star actually ends up slightly cooler as explained in Section 2.3 (Figure 1.3). The nuclear reaction rates are sensitive to the temperature, and so the nuclear burning slows as the expansion occurs and the temperature drops. The net effect is a highly efficient process of negative feedback. If the star temporarily produces an iota too much heat, the nuclear fires are automatically damped a bit by the expansion to restore equilibrium. The opposite is also true. If the nuclear burning should fail to keep up with the energy radiated away for an instant, the heat would be insufficient, the pressure would drop, the star would contract, and the temperature would rise. The result is that the nuclear burning would be increased to the equilibrium value. A star burning hydrogen on the main sequence thus works in a manner similar to the thermostat and furnace in a house. If the temperature drops, the furnace kicks in and restores the lost energy. If the house gets too hot, the furnace turns off temporarily until the desired temperature is restored.

The process of hydrogen burning on the main sequence is one of thermonuclear fusion. Nuclei of hydrogen atoms, protons, are fused together to make the nucleus of the heavier element helium, which consists of two protons and two neutrons. Burning hydrogen to helium depends primarily on the nuclear force. The role of the nuclear force is to bind the four particles in the helium nucleus. The energy left over from combining the particles is available as heat. This process is not different in principle from ordinary burning where chemical forces bind the combined products together and liberate the energy of combining the molecules as heat. Chemical forces are based on the electrical force. The reason that nuclear burning is so much more powerful than chemical burning is because the nuclear force is so much stronger than the electrical force. The energy released in the fusion of hydrogen into helium is an appreciable fraction, about 1 percent, of the maximum amount of energy that could be released if all the mass of hydrogen were turned into pure energy in accordance with $E = mc^2$. That very high efficiency of energy release is why thermonuclear bombs are such a fearful weapon and why the promise of controlled thermonuclear fusion is so enticing as an ultimate energy source.

Look more closely at the process of turning hydrogen into helium. There are many ways in which this can be done in practice, but they all have a common link. The process of thermonuclear fusion consists of combining four protons to make helium. Of necessity, some step in this process requires that two of the protons be converted into two neutrons. Protons are converted into neutrons (and vice versa) by the influence of the weak force. To understand how this process works, and to reveal an important practical consequence, we must also invoke the laws of conservation of charge and of baryons and leptons as introduced in Section 2.2.

The conversion of two protons into two neutrons during hydrogen fusion conserves the number of heavy, baryon, particles; there are two to start and two in the end. That process cannot occur alone, however, because charge is not conserved; the charge on the protons cannot just disappear. One way to get around this is to produce two positively charged particles to balance the charge on the protons and to give no net change in the electrical charge. These positive particles cannot be baryons of any kind because the number of baryons in the reaction is already balanced. Nature solves this problem by providing leptons in the form of positrons. If two protons are converted into two neutrons and two positrons by the weak force, we have no net charge. Now, however, we are making two new leptons, and to conserve the lepton number, the reaction must spit out two other leptons along with the two neutrons. Recall from Section 2.2 that positrons have the opposite charge and the opposite leptonness from electrons. Algebraically, they each count as "minus one" lepton in the exit channel. The other leptons coming out of the reaction must carry no charge, because the charge is already properly balanced, but must count as "plus one" in terms of leptons in order to offset the positrons. To balance charge, baryons, and leptons all at once in this reaction, nature provides the neutrino!

The fact that the neutrino was needed to conserve all the relevant quantities in certain nuclear reactions was first realized by the Italian physicist, Enrico Fermi. It was Fermi who gave the particle its name, meaning little neutral one. Fermi was awarded the Nobel Prize for this and related work in 1938 as he prepared the world's first nuclear reactor and took seminal steps that would lead to the Manhattan Project in World War II. The neutrino was not directly detected until after the war in the 1950s when Fred Reines and colleagues registered neutrinos coming from a nuclear reactor. Reines was given the Nobel Prize for this discovery in 1995.

Figure 1.6 summarizes the essential processes that occur when hydrogen undergoes thermonuclear fusion to make helium. In that conversion, a neutrino must be made for every neutron that is produced in order to conserve baryons, leptons, and electrical charge simultaneously. For every atom of helium produced, two neutrinos must be generated. That fact represents both an opportunity and a challenge to astronomers and physicists.

3.3. The Solar Neutrino Problem

Hydrogen burns and neutrinos are produced in the centers of stars because that is where the temperature is the highest. Because neutrinos interact only by the weak force, normal stellar matter is virtually transparent to them. The neutrinos that are produced in the central hydrogen-burning reactions immediately flow out of the star at nearly the speed of light, as shown in Figure 1.7. They carry off a small

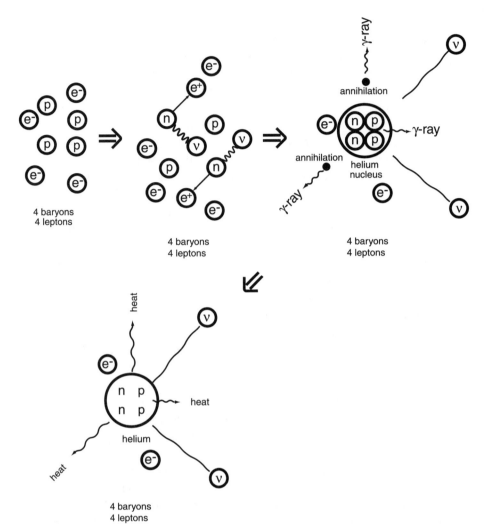

Figure 1.6. The process of hydrogen burning involves the thermonuclear fusion of hydrogen to make helium: (top left) the original hydrogen gas consists of equal numbers of protons and electrons, four baryons and four leptons; (top middle) under the combined action of the strong and weak forces, two of the protons are converted to two neutrons plus two positrons and two neutrinos. The net electrical charge is still zero and because positrons represent antileptons, there are still only four baryons and a net of four leptons; (top right) the strong force binds the two remaining protons and the two newly created neutrons into a nucleus of helium. This process releases a large amount of heat in the form of the radiant energy of gamma rays. The positrons annihilate upon collision with two of the initial electrons and produce a little more gamma-ray energy. The net result is still four baryons – two protons and two neutrons – and four leptons – two of the remaining initial electrons and two newly made neutrinos; (bottom) the final product is a new helium nucleus, heat, and two neutrinos that race out of the star and into space.

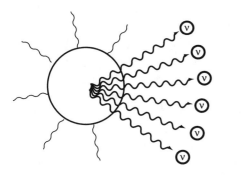
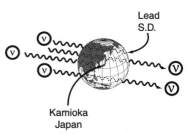

Figure 1.7. Neutrinos produced in the thermonuclear burning of hydrogen to helium in the center of the Sun flood into space. Some of the neutrinos head in the direction of the Earth. Most of the neutrinos that reach the Earth also pass right through it, but a few can be stopped and studied in special detectors. The pioneering solar neutrino detector was in Lead, South Dakota. The currently most successful is in Kamioka, Japan.

amount of energy that would otherwise be available to heat the star, but this energy is not of great import. The importance of the neutrinos to astronomers is that they come directly from the center of the star, carrying information about conditions in the stellar core. Otherwise, astronomers are limited to studying photons of light that come only from the outer surface of the stars. Study of these photons is a powerful tool to deduce the nature of the inner portions of a star, but that is no substitute for being able to directly "see" inside. Neutrinos from the Sun provide that opportunity.

The problem with observing the heart of the Sun by means of neutrinos is that the neutrinos will stream through any detector unimpeded, for the same reason that they stream freely out of the star. Detection of the neutrinos depends on amassing a huge detector and then waiting for that rare time when the weak force causes a reaction within the detector. This process is totally impractical for any star but the Sun, because the great distance dilutes the neutrino "brightness" from a distant star as rapidly as it does visible photons.

The first successful effort to detect neutrinos from the Sun was the result of a multidecade effort by Ray Davis and his collaborators (see Figure 1.7). This work has not yet won a Nobel Prize, but it should. The detector consists of a hundred thousand gallons of chlorine-rich cleaning fluid. The chlorine undergoes an interaction with a neutrino by means of the weak force. This interaction turns a neutron within a chlorine nucleus into a proton, just the opposite of the reaction that produced the neutrino in the Sun. Changing a neutron in chlorine into a proton converts an atom of chlorine into an atom of radioactive argon. The argon can be collected efficiently because it is a noble gas and does not combine chemically. The tank containing the cleaning fluid is at the bottom of the Homestake gold mine in Lead, South Dakota. The underground operation is necessary to screen out cosmic ray particles that could induce spurious transitions of the chlorine to argon. The mine was vacant until the price of gold soared to astronomical highs several years ago. The Homestake company reactivated it, and for a while the scientists had to work to the sound of dynamite explosions as new veins were developed.

At first, the solar neutrino experiment gave no signal at all above the background "noise" of extraneous reactions. This caused a great deal of anguish in the astronomical community because the first opportunity to peer directly inside the Sun gave a result inconsistent with apparently straightforward theoretical predictions. With patience, a positive signal was detected. A few hundred atoms of argon are collected each month from the hundred thousand gallons of fluid! Detection of some neutrinos is more reassuring than detection of none at all, but a new serious problem still arose. The most careful analysis of a standard computer model of the Sun predicts several times more neutrinos than are observed.

The discrepancy could lie in several areas. The nuclear reactions could proceed in a different manner than we envisage. The structure of the Sun could be somehow different. Perhaps the composition, particularly the heavy elements, is not spread uniformly through the volume, as assumed. Perhaps the fundamental properties of the neutrinos themselves are different. The gold mine experiment is looking for the particular type of neutrino produced when protons change to neutrons. There are (at least) two other kinds of neutrinos. If the neutrinos have undergone a Jekyll and Hyde transformation in flight and are one of the other types when they arrive at Earth, they would not induce the desired transformation of chlorine to argon and would go undetected.

Recent developments may have given the key to this mystery. One reassuring result came from an underground neutrino detector constructed in Kamioka, Japan, called Kamiokande (Figure 1.7). This detector is a massive vat of water. Unlike the chlorine experiment, it can see neutrinos in real time and can tell the direction in which the neutrinos are moving and hence the direction from which they came. The neutrinos can trigger the conversion of a neutron to a proton in the oxygen in the water or collide with one of the electrons in the water. In either case, the particle that is hit is given substantial energy and flies rapidly through the water in the direction that the neutrino was traveling. The recoil particles give a flash of blue light known as Cerenkov radiation in the direction in which they are moving. From this flare of light in the detector, the direction of the neutrinos can be tracked. The Kamiokande experiment saw the same kind of neutrinos as the chlorine experiment and at the same low rate, but, to everyone's great relief, the neutrinos were definitely coming from the direction of the Sun! Without that confirmation, there was a small probability that the Homestake detection was some local contamination and not solar neutrinos at all. That would have made the problem even worse.

The second development may have given the real answer. The Homestake and Kamiokande experiments detect only the stream of the few high-energy, relatively easy to detect neutrinos that come from a rare version of the hydrogen-burning process. That rare process might be affected by subtle changes in the interior of the Sun that would not affect the overall power output. The chlorine and water experiments cannot detect the far more numerous neutrinos that must be produced in the basic reaction by which a proton is turned into a neutron at a rate that is directly proportional to the power that flows in radiation from the surface of the Sun. Another experiment, carefully planned for a decade in collaboration between Ray Davis and Russian physicists, uses the element gallium as a detector. This substance is sensitive to the basic flood of low-energy neutrinos that must be there because the Sun, after all, is shining. The gallium experiment also failed to see the predicted rate of neutrinos! The only remaining conclusion is that something is omitted from our simplest physical picture of the neutrinos.

As mentioned earlier, there are three different types of neutrinos, each with their antineutrinos. That there are three types of neutrinos is related to the fact that there are three types of quarks that make up other particles like protons and neutrons. When neutrinos were first discovered, it was suspected that they had no mass. If that were the case, each type of neutrino would always be the same. The fledgling grand unified theory combining the strong and electroweak forces suggests that neutrinos must have a small mass. In that case, the theory predicts, there are circumstances in which one type of neutrino can be converted to another type. If this happens round and round and back and forth among the three types of neutrinos, then by the time the neutrinos arrive at the Earth there might be roughly equal amounts of all three. In this case, only one-third of the type originally produced in the Sun that the experiments were specifically designed to register would reach the detectors. The fact that about one-third of the expected rate is observed is consistent with this notion.

This interpretation of the solar neutrino experiments strongly suggests that we not only have at last the solution to the solar neutrino problem but also have strong evidence for the grand unified theory of elementary particles. This is probably the answer, but it also raises the challenge of building more experiments to test the hypothesis.

A major step in this saga was announced in the summer of 1998 by the teams of scientists working on the new, larger underground experiment in Japan known as Super Kamiokande. This experiment found evidence that neutrinos do shift from one type to another as they interact with the Earth's atmosphere, and hence that they must have a mass, as expected from theory. The mass is not measured directly, only the difference in the masses, but this is a major breakthrough. On the other hand, to account for all the data from all the experiments, there is some discussion of the need to introduce yet another type of neutrino called a "sterile" neutrino that interacts only with neutrinos and with no other particles at all. This seems a step backward. Study of solar neutrinos still has much to teach us. We will return to neutrinos in another context in Chapters 6 and 7.

2

Stellar Death

The Inexorable Grip of Gravity

1. Red Giants

The Sun looks the same to us, unchanging, day after day. A simple observation, however, tells us that it is evolving and must be changing in some manner. That observation is just the warmth on our upturned faces on a sunny day. The radiation that flows from the Sun carries energy out into space. There is nothing from space replacing that energy. The Sun must, therefore, be losing energy overall. Something must be going on within the Sun that is slowly, inevitably altering it. The lesson from Chapter 1 is that the change in the Sun involves its composition. The Sun is irrevocably transmuting some of its hydrogen into helium. That transformation cannot be undone. The alteration of the structure of the Sun is slow, but it is steady. Eventually, the changes will be drastic.

As remarked in Chapter 1, the hydrogen burns only in the center of a star, where the temperatures are highest. That means that the central region is where the hydrogen is consumed and the helium builds up. Even when the hydrogen is fully transformed in the central region, the outer, cooler portions of the star will not have burned. They retain their original composition. This causes the star to become schizoid and to do two things simultaneously: shrink and swell. This development is in strict accord with the principle of conservation of energy, but the application of this principle is more complex than for stars with a homogeneous composition.

When hydrogen is exhausted in the center, the star has a central volume of nearly pure helium (along with the scattering of heavy elements initially present in the star). The remainder of the star is original material, composed mostly of hydrogen. The difference between the inner parts of the star where the composition has been altered and the outer part where the composition is unchanged become ever more distinct as the star evolves. To distinguish these two portions of the star, the inner part is called the *core*, and the outer part, the *envelope*.

The helium in the stellar core can become a thermonuclear fuel. Helium burning does not happen spontaneously, however, any more than hydrogen burning did. The nuclear force is strong, but it only acts over very short distances. The particles to be combined must be brought close together. There is, however, a force that inhibits the particles from getting close to one another. This is just the electrical force of the repulsion of like charges. The nuclei of atoms, such as hydrogen, helium, or heavier elements, are composed of positively charged protons and neutrons with no electrical charge. All nuclei thus have a net positive charge. If the electromagnetic force and gravity were the only forces in the Universe, this charge repulsion would prevail, and there would never be any nuclear reactions.

To initiate thermonuclear burning, the charge repulsion among the protons must be overcome. The electrical repulsion is not as strong as the nuclear force, but it acts over greater distances and dominates while the particles are far apart. At close distances, the nuclear force is stronger, and it can grab the particles and bind them tightly together. To bring like-charged particles together so the nuclear force can grab them and liberate energy, some energy must first be expended to fling the particles together despite the resistance of the electrical repulsion. You do not get something for nothing, but the nuclear payoff is worth the investment of some energy to overcome the charge repulsion. This principle is illustrated in Figure 2.1.

In practice, the charge repulsion is overcome by investing the particles with heat energy. This gives them more random energy of motion so they collide more fiercely and come closer within the grasp of the nuclear force during an encounter. To burn a nuclear fuel, you have to heat it first by raising the temperature, just as you need a match and kindling for the wood in a fireplace. A protostar must

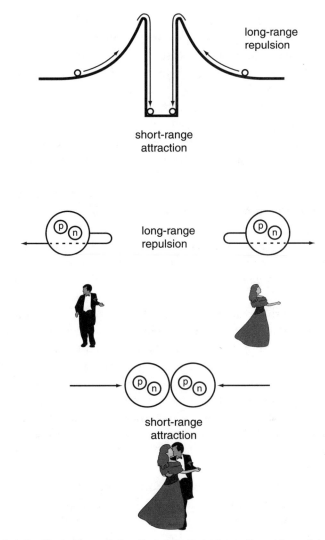

Figure 2.1. Positively charged atomic nuclei repel one another at long distances but are strongly attracted at short distances by the nuclear force. In analogy to a deep hole surrounded by a raised lip (a "volcano"), some energy must be invested as heat to force the nuclei close to one another or to roll a ball up the hill. After the nuclei are sufficiently close together, the short-range nuclear force can bind the nuclei together to make a new element and liberate energy, just as a ball, having reached the precipice, can plunge down into the crater, yielding more energy than it took to roll it up the hill. In practice, the atomic nuclei need to have some neutrons so that their nuclear attraction can overcome the charge repulsion of the protons.

contract sufficiently to heat the hydrogen to get burning started initially. Helium nuclei have two protons, whereas hydrogen nuclei have only one. The charge repulsion is stronger for helium than for hydrogen, so helium must be heated to higher temperatures than hydrogen before it undergoes thermonuclear reactions.

When the last of the hydrogen burns out in the center of a star, the star must get even hotter to burn helium. It solves this problem in a natural way, using energy conservation. After the hydrogen burns out in the center, no energy is being produced. Without the input of heat, the pressure cannot support the star. The star thus contracts and derives heat that way until the helium becomes hot enough to burn. The same mechanism that is responsible for igniting and regulating hydrogen burning on the main sequence causes the helium to ignite after the hydrogen is exhausted in the center. When the star has insufficient heat, it naturally contracts until that heat can be provided, whether by hydrogen, helium, or ultimately other sources of nuclear fuel.

Now comes the schizophrenia. The helium core contracts and heats until helium ignites. In its inimitable way, the gravitational contraction liberates more heat energy in the core than the core needs. The excess heat flows out into the overlying envelope of pristine material. The envelope responds in its own natural, but opposite, way. The envelope feels that it is getting an excess input of heat. The excess pressure causes the envelope to expand against gravity and cool to lower temperatures. The star thus does both things at once. The core loses energy, contracts, and heats, and the envelope gains energy, expands, and cools.

The contracting core is more important for the ultimate evolution of the star, but what astronomers actually see in their telescopes is the outside of the envelope. The outside, like the whole envelope, gets cooler and hence more red in color. Inside, the helium burns at a high rate and provides a high brightness for the star. At a given surface temperature, astronomers categorize the brightest stars as giants and the rather dim stars as dwarfs. The stars we are describing have become what astronomers call *red giants*. The size of such stars also becomes very large as the envelope expands, so the star is also a giant in terms of its extent, even though this is not technically what an astronomer means by giant. For instance, a blue supergiant is much brighter, but much smaller than a red giant. In any case, red-giant stars swell from the size of the Sun to extend well beyond the radii of the inner planets of the Solar System. We expect the Sun to undergo this transition in about another 5 billion years, at which time the inner planets should be engulfed and evaporated. The Sun will live about 1 billion years, about 10 percent of its total lifetime, as a red giant and then die.

To be fair, this explanation for the formation of a red giant by exchange of energy from the core to the envelope is a little simplistic. The exchange of energy does happen and is one factor, but experts still argue about the best way to understand why red giants form. The computer models show that it happens, but the process is a complex, nonlinear interaction of the star with gravity and is not that susceptible to simple, this-is-the-key-factor-type explanations. In a certain sense, the formation of a red giant involves an instability. It is as if you push a book toward the edge of a table. Nothing much happens for quite a while. If you push too far, however, the book will land on the floor. As the core shrinks in a star that has consumed its central hydrogen, there comes a point where the envelope "falls" outward, coming to a lower energy solution that couples the pressure in the core and envelope with gravity.

Stars with appreciable mass pass through several burning stages after they become red giants. They also spend about 10 percent of their total life in this phase, with each stage progressing faster than the one before. After each successive fuel is exhausted in the center, the star finds itself without a source of heat, so the core contracts until the material that was formed by the previous burning phase becomes hot enough to burn. The core must become hot enough to overcome the charge repulsion among the greater number of protons in ever more complex nuclei. In massive stars, hydrogen burns to become helium in the basic way we described in Chapter 1. The details are different than those for the Sun, but the net outcome is the same: four protons must combine to make a helium nucleus with the creation of two neutrinos.

In stars with the mass of the Sun and in more massive stars, helium burns to become carbon and then oxygen. The reason for this is that the simplest interaction one can imagine, combining two helium nuclei, makes a nucleus with four protons and four neutrons. For reasons that have to do with the details of how the nuclear force works, the nuclear attraction of that combination of protons and neutrons is not able to overcome the charge repulsion of the four protons. The combination of four protons and four neutrons is unstable. A nucleus with four protons and four neutrons falls apart and hence cannot be one of the steps in nuclear burning to produce a heavier "ash" from a given fuel.

Nature finds a way around this bottleneck by utilizing the more rare process by which three helium nuclei occasionally become close enough to combine under the control of the nuclear force. The result is a nucleus with six protons and six neutrons, the element carbon! This is where all the carbon necessary for life arises. As the helium burns in this way, some of the as yet unconsumed helium can combine with the newly formed carbon to make an element with eight protons and eight neutrons, the element oxygen, another critical agent for life as we know it.

In the Sun, thermonuclear burning is expected to halt with the production of carbon and oxygen for reasons that will be addressed in Section 3. For sufficiently massive stars, the process continues. Ultimately, a complex of heavier elements forms. Prominent among these substances are the elements neon, magnesium, silicon, sulfur, argon, calcium, and titanium. That may seem an odd assortment, from a noble gas to the stuff in your bones to a metal used in submarine hulls, but there is a common factor. Each of those successive elements consists of two more protons and two more neutrons than the one before. Stars produce this chain of elements in especially large abundance because each is essentially made up of the basic building blocks of helium nuclei: three for carbon, five for neon, and ten for calcium. Each successive element contains more protons than the last because each phase of burning is one of fusing lighter nuclei into heavier ones. More protons means more charge repulsion to be overcome by higher temperatures. The star obligingly provides the higher temperature in the core by contracting whenever it finds itself without any nuclear energy input to balance the radiation energy lost to space.

This seductive process by which a star prolongs its life actually just puts it deeper and deeper in the grip of gravity. Gravity will ultimately win the battle.

2. Stellar Winds

Before delving into the depths of the stellar cores, let us consider some of the important processes in the outer parts of the star by which some stellar matter can escape the grip of gravity.

On the Earth, a wind is the actual motion of matter, air molecules moving en masse from one place to another. In addition to radiation, the Sun emits a wind of particles, mostly hydrogen, that flows out into space in all directions. For the Sun, the cause is not precisely known. It may be due to the turbulent, boiling surface pumping magnetic energy into the outer layers and expelling them. Evidence for the solar wind is in the tails of comets. Comet tails always point away from the Sun, wafted by the stellar breeze, whether the comet is headed toward or away from the Sun. The solar wind is interesting, but the total amount of matter expected to be lost from the Sun during its lifetime on the main sequence is negligible. The nature of a wind from a star is illustrated schematically in Figure 2.2.

For more massive stars, the story is different because the loss of mass to a wind can substantially alter the evolution of the star. For massive stars, the mechanism to expel matter is thought to be the pressure of the intense radiation that flows from the star. Although we turn to the Sun for warmth, we do not usually think of the pressure of the sunlight on our faces. It is there, but it is very small. In space with no competing effects, the pressure exerted by the photons of radiation streaming out from the Sun can be appreciable. There are dreams to have a sail plane race in space with all the craft powered by the pressure of the solar radiation.

The power emitted in radiation from a star is known as the star's *luminosity*. The luminosity is the amount of radiation energy that flows from a star in a given time. The pressure exerted by the radiation is proportional to the luminosity. As the mass of a star goes up, the luminosity and the pressure exerted by the radiation increase by about the third power. That means that if you consider a star of twice the mass, the luminosity goes up by a factor of eight. For a sufficiently large stellar mass, the large radiation pressure associated with the large luminosity becomes a dominant process. In massive, bright stars, the pressure of the radiation flow is much greater than it is for the Sun. For massive stars, the radiation pressure in the outer parts of the star can be so great that matter is actually blown off the surface of the star in appreciable quantities. This is thought to be the mechanism behind the large stellar winds from massive stars.

Because of the very strong stellar winds, massive stars can lose a large part of their mass while they slowly burn hydrogen on the main sequence. After a massive star leaves the main sequence, the lifetime gets shorter, but the rate of loss of mass in a wind is much higher. The result is that appreciable mass can be lost in the red giant phase

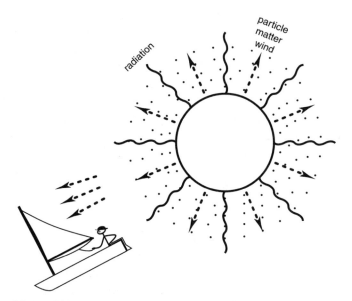

Figure 2.2. In addition to the flow of radiation from the surface of the Sun or other stars, there can also be a flow of matter, a stellar wind.

even if relatively little has been shed on the main sequence. Large mass loss can affect the evolution of the star. If the wind is strong enough, the entire hydrogen envelope can be expelled, thus exposing the core of helium and heavier elements.

Stars with less than about 30 solar masses can lose enough mass in a wind that they end up with substantially less mass than they had when they were born. This does not affect the qualitative behavior of the star, but it can alter details of the evolution. Stars of this relatively low mass do not have sufficiently strong winds to expose the core. In some cases, however, a binary stellar companion can tug the outer mass off and still produce a bare core with little or no hydrogen blanket. This and other effects of binary companions will be discussed in Chapter 3. Stars with mass between about 30 and 50 solar masses do become red giants but then are thought to undergo such an appreciable loss of mass to a stellar wind that the red giant envelope is ejected anyway, exposing the core. For stars in excess of about 50 solar masses, there is no observed red-giant phase. The interpretation is that so much mass is lost on the main sequence due to a strong stellar wind that no outer hydrogen envelope is left to expand and become a red giant.

If the entire hydrogen envelope is lost to a wind, the bare core composed of helium and heavier elements should be exposed to view. We observe stars of just these properties. The *Wolf-Rayet stars* have little or no hydrogen on their surfaces and are seen to have strong winds themselves. Wolf-Rayet stars are thus thought to be the result of strong mass loss by winds from massive stars. This means that massive stars may not be red giants when they undergo core collapse but rather Wolf-Rayet stars. Whether Wolf-Rayet stars explode as supernovae or collapse to make black holes or some mix of both is not known.

As already noted, radiation pressure exerted by a star is proportional to its luminosity. There is a critical luminosity above which the outward radiation pressure exceeds the inward pull of gravity. In this case, the result is not just a wind but rather a complete disruption of the balance of pressure and gravity in the star. This limit to the luminosity is called the *Eddington limit*, after the early British astrophysicist, Sir Arthur Eddington, who first realized the key role radiation could play in stars. The critical Eddington limit luminosity is proportional to the mass of the gravitating star; it is the gravity of that mass that the radiation pressure must overcome. A star of fifty times the mass of the Sun is so bright that it is near the Eddington limit. That is why it blows such a substantial wind.

In Chapters 5, 8, and 10, we will also talk about circumstances when matter is dropped onto a compact, high-gravity star, like a white dwarf, a neutron star, or a black hole. Radiation pressure can also play a crucial role in these circumstances. If matter falls onto a star of high gravity, a great deal of heat and luminosity are generated. The resulting luminosity can exceed the Eddington limit, and the associated radiation pressure can actually prevent matter from falling onto the star at any higher rate. If the rate were higher, the excess matter would be blown away rather than falling on the star. The rate of infall of mass that just provides the Eddington luminosity is known as the *Eddington mass accretion rate*. In principle, a balance can be achieved in which the radiation pressure allows only enough mass to fall onto a compact star to generate the Eddington limit luminosity that provides the pressure. A star in such a balance will automatically radiate precisely the Eddington limit luminosity and the mass infall onto it will be precisely the Eddington mass accretion rate.

3. Quantum Deregulation

Let us now return to what happens in the guts of a star as it evolves. Section 1 described thermonuclear burning in conditions where the thermal pressure dominated over the quantum pressure. In this situation, the star can regulate its burning because the star will heat up when it loses energy and cool off if it gains energy. The process of contracting and heating and passing from burning phase to burning phase is halted if the core of the star gets too dense. At high density, the electrons are squeezed together so much that the exclusion and uncertainty principles comes into play as described in Chapter 1. In this circumstance, the quantum pressure of the electrons exceeds the thermal pressure of the electrons and atomic nuclei. This happens first for lower-mass stars that are denser than high-mass stars at a given burning phase.

In this compact state governed by the quantum pressure, the star loses the ability to heat and ignite a new, heavier nuclear fuel. Any nuclear fuel that does burn under these conditions is not regulated. The star loses the ability to control its burning and its temperature. The quantum pressure deregulates the temperature; the thermostat of the star is broken.

The reason for this quantum deregulation is that the quantum pressure does not depend on temperature. If the star supported by the quantum pressure loses a net amount of energy because the nuclear fires have gone out, the pressure remains unchanged. There is no contraction to provide heat, so the temperature just drops as the heat is lost. A star, or portion of a star supported by the quantum pressure, behaves as you would think normal matter should: when it radiates away heat, it cools off, as illustrated in Figure 1.3. If a nuclear fuel ignites in a star supported by quantum pressure, the burning adds some heat. The pressure does not rise, so there is no expansion to absorb the heat. The temperature simply rises. The nuclear burning is very sensitive to the temperature, however. Thus at the new higher temperature, the burning proceeds even faster, raising the temperature even more. The nuclear rates can become so fast that the energy they produce can blow the star to smithereens. A star supported by the quantum pressure has an unstable, unregulated temperature. The temperature will decline toward absolute zero if there is no nuclear burning. The temperature will rise sharply if there is nuclear burning. The star has a broken thermostat. Even more, it is as if, when your house gets a little hot, you set the rafters on fire. The way in which the quantum deregulation sets stellar rafters aflame is given in Chapter 6.

Most stars reach this state of unregulated temperature and burning after helium has burned out in the core. The core is then composed of a mixture of carbon and oxygen. The core typically has a mass about 60 percent of the mass of the Sun, independent of the total mass of the star. This applies to all stars with mass up to about ten times the mass of the Sun, and that is most of the stars. The remaining mass is in the extended red-giant envelope. While the envelope is as big as the Earth's orbit, the core is very tiny by the time the quantum pressure becomes dominant – a few thousand kilometers in diameter, about the size of the Earth. The resultant density can be a million to a billion grams per cubic centimeter. Ordinary earthly matter, or that in normal stars, is about one gram per cubic centimeter. To get such high densities that the quantum pressure comes into play, a whole building, such as the seventeen-floor physics building in which I work, would have to be packed into the volume of a sugar cube. Only gigantic gravitational forces can achieve such a compaction.

This small dense core is immediately surrounded by two narrow, very bright shells of matter where helium and hydrogen are burning. These shells are the last remnants of the stages of hydrogen and helium burning in the center of the core through which the star has already passed. The pressure of radiation from these burning shells causes the envelope to pulsate violently and blow matter from the star. The outer envelope is ejected in this process. Astronomers see the outcome of this process as a shell of gas proceeding outward from the star. These expanding, ejected shells are called *planetary nebulae*. They have nothing to do with planets except that they are often sufficiently extended in photographs that, like planets, they do not have a "starlike" point image. Planetary nebulae were misnamed by early astronomers, but the name has stuck.

When the envelope is ejected, the core of the star is left behind. Supported by the quantum pressure of its squeezed electrons, the core cools off to become what is known as a *white dwarf*. When a white dwarf forms, it still has a great deal of heat and looks blue-white to an astronomer. The term "dwarf" comes from the low luminosity. The white dwarf has such a small surface area that the white dwarf is dim despite its high temperature. White dwarfs are also tiny in size and hence dwarflike in that sense, even though, again, that is not the meaning astronomers have attached to the word. We will return to white dwarfs in Chapter 5.

4. Core Collapse

Massive stars continue to evolve, forming cores within cores of ever heavier elements until the innermost regions are turned into iron. Iron is a very special element in the Universe. It is almost composed of fourteen helium nuclei but is a little more complex because two of the protons have converted to neutrons, so iron has four more neutrons than protons. By the happenstance of the nature of the strong nuclear force among protons and neutrons, the fifty-six particles of an iron nucleus are more tightly bound together than in any other element (with the possible exception of a couple of exotic elements like rare isotopes of nickel, which cannot easily be formed in nature). Iron happens to be at the bottom of a nuclear "valley" toward which all other elements would like to fall, just as rocks roll down a mountainside, as shown in Figure 2.3. The difference is that the force causing the settling toward the "bottom" is the nuclear force, not gravity. All

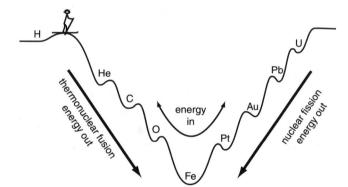

Figure 2.3. The element iron sits at the bottom of the nuclear "valley" defined by the nuclear and electromagnetic forces. Light elements, shown here schematically as hydrogen, helium, carbon, and oxygen, need to be heated to overcome the "bumps" representing charge repulsion, but then they can fuse into heavier elements, end up deeper in the valley, and thereby release a net amount of energy. Heavier elements, shown here schematically as platinum, gold, lead, and uranium, will liberate energy, slipping deeper into the valley, if they fission into lighter elements. Iron can be transmuted only by putting energy into it, either to break it apart into lighter elements or to fuse it into heavier elements. The result is that iron can only absorb energy from a star, never produce energy.

elements lighter than iron would energetically prefer to merge together to form iron. They are prevented from doing so only by the repulsion of the electric charge on the protons. Stars are nature's way of overcoming the electrical repulsion and rolling the elements down the nuclear hillside to the bottom where iron comfortably sits.

As rocks roll downhill, they turn their gravitational energy into other forms, such as noise, breaking trees, dislodging other rocks, and compacting and heating the soil where they land. This complex process conserves the total energy. When the rock is at the bottom of the valley, it can roll no farther, and no more energy can be obtained from it. A similar process occurs in forming iron. Energy is released as light elements fuse together to form heavier ones closer to iron. Elements heavier than iron are on the other side of the valley from the light elements, but their protons and neutrons are also less tightly bound than those of iron. These elements approach iron by splitting apart into lighter elements in the process called *nuclear fission*. This process is the one that powers nuclear reactors, but it does not occur naturally in stars to any great extent because the stars are composed of elements lighter than iron.

Energy cannot be obtained from a rock at the bottom of a valley. On the contrary, to move the rock, energy must be invested to lift or roll the rock back up the hillside from which it originally fell. What about a stellar core made of iron? No more nuclear energy can be derived from that core. With no nuclear energy input, the star radiates a net amount of energy into space. The massive stars that develop iron cores are typically hot enough that thermal, not quantum, pressure dominates their structure. Thus when such stars lose energy, gravity squeezes them, and they heat up. Gravity naturally makes energy available to the iron. The response of the iron is to roll up the nuclear hillside. Most of it breaks apart into the lighter nuclei from which it originally formed. Some of the iron will undergo fusion reactions that lead to the heavier particles on the other side of the valley. Both of these processes require energy. Rather than firing up a new nuclear reaction to repel the squeeze of gravity, the iron absorbs heat energy from the star. The hot particles exerted the thermal pressure to support the star. When the particles lose energy to the breakup of iron, the pressure cannot rise. Gravity then compresses the iron core even more, but the iron continues to break apart, absorbing the energy and preventing a rise in pressure to withstand the stronger gravity.

The result is another example of energy conservation, with iron playing the negative role of a sponge rather than a source of energy. With iron absorbing energy, gravity overwhelms the weakened pressure. The formation of an iron core in a massive star signals the end of the thermonuclear life of the star. At that point, the star is doomed. Gravity prepares to deal the death blow. The core collapses in a mighty implosion!

5. Transfiguration

As the iron disintegrates into lighter elements in the collapse, the core plunges to a smaller size, and the density skyrockets. The rising quantum pressure of the electrons is too feeble. The electrons stop fighting the gravity and disappear. They do this by combining with a convenient proton (a mutual suicide pact determined by the conservation of charge) and forming a neutron. To conserve lepton number, a neutrino must be produced for every electron that disappears, as discussed in Chapter 1. The result is that in the collapse of the iron core, the electrons and protons disappear to be replaced by neutrons and a flood of neutrinos. The result is the formation of an entirely new type of astronomical object, a *neutron star*.

A neutron star is composed almost entirely of neutrons. The mass of a typical neutron star is somewhat more than that of the Sun, and its radius is about 10 kilometers. This is only about the size of a small city like Austin. The density at the center of a neutron star exceeds that in the nucleus of an atom. In a sense, a neutron star is a gigantic atomic nucleus held together by gravity. A typical density would be about 10^{14} grams per cubic centimeter. To attain such a density, an entire city like Austin would have to be packed into the size of a sugar cube.

The gravity of a neutron star is fantastically large and must be balanced by an equally large pressure. At a large enough density, the quantum pressure of the neutrons can become sufficiently great to overcome the force of gravity and restore the condition of dynamic equilibrium. The quantum pressure of the neutrons is aided by the nuclear force. As described in Section 1, the nuclear force has no effect on particles that are a large distance apart; however, when they get quite near, the nuclear force pulls them together. The nuclear force is "attractive," like gravity or opposite charges. An important detail mentioned in Chapter 1 comes into play when nuclear particles are packed very close together. At very small distances between particles, the nuclear force drives baryons apart. The nuclear force becomes "repulsive," like similar charges. This repulsive force on closely packed neutrons helps to hold them apart and contributes to the pressure that supports a neutron star. As for white dwarfs, there is a maximum mass to neutron stars, a maximum mass that can be supported by the combined quantum and nuclear pressure of neutrons. The quantum effects are known precisely, but the nuclear force is not exactly established, so this pressure, and hence the total mass it can support, are still somewhat uncertain. The best guesses based on sophisticated calculations of nuclear matter are that the maximum mass for a neutron star is of order 1.5–2 solar masses.

The process of collapse and renewed support by the quantum pressure of the neutrons and the repulsive nuclear force among very compact neutrons is quite rapid. It requires only about a second in a star that has lived for millions and millions of years in tranquillity. The details of this process will be explored in Chapters 6 and 7. A summary of what we have learned about neutron stars will be given in Chapter 8.

There is no guarantee that the process of core collapse will result in the formation of a neutron star. A tremendous amount of gravitational energy is released in the collapse, a hundred times more energy than is necessary to blow the outer layers away from the star. One reason that the nature of neutrinos was stressed in Chapter 1 is that they play a dominant role in core collapse. The majority of gravitational energy produced in the creation of a neutron star, more than 99 percent, is given to the neutrinos. The neutrinos escape from the collapsing iron core and the newly formed neutron star and carry most of the energy off into space.

The degree to which the remaining energy available from collapse may be directed outward rather than inward is not clear. If a fraction of the energy is used to blow off the layers of the star surrounding the original iron core, then a neutron star can be left behind. On the other hand, if insufficient energy is directed outward to eject the outer portions of the star, then the outer layers rain inward. A neutron star may form momentarily from the collapsed iron core, but then the rest of the star falls inward. Because we are talking about a process that occurs in massive stars, the mass that falls in will far exceed the maximum mass a neutron star can support. The neutron star will rapidly be crushed out of existence in a process of total, ultimate collapse. The result will be the unique gravitational entity that astrophysicists call a *black hole*. A black hole is an object for which all the mass has been crushed to what is effectively zero volume. All that remains is the gravitational field that becomes overwhelming at distances close to the center of the collapse. We will study the details of these fantastic objects in Chapters 9 and 10.

3

Dancing with Stars

Binary Stellar Evolution

1. Multiple Stars

Cecelia Payne-Gaposhkin was a pioneer of modern astronomy. She devoted much of her research to the study of multiple star systems and coined a comic adage to describe one of the basic tenets of that work: "Three out of every two stars are in a binary system." By this she meant to illustrate that roughly half the stars in the sky have companion stars in orbit. If you were to look closely at half the stars you would find that there are two stars where a more casual examination would have revealed only one point of light. Many people know that the nearest star to the Sun is Alpha Centauri. Less well known is that Alpha Centauri has a companion in wide orbit, known as Proxima Centauri. A closer examination shows that Alpha Centauri itself is not a single star but has a closely orbiting companion as well. Of the "two" stars closest to the Sun, three are in the same mutually orbiting stellar system.

Stars occur in many combinations. Single stars and pairs are most common, but some systems contain four and five stars in mutual orbit. In this chapter, we will concentrate on the systems with a pair of stars, double stars, or, somewhat more technically, binary stars (but we try to refer to the phenomenon of duplicity, not the mangled–jargon born "binarity" that creeps into the literature). Binary stars come in two basic classes: wide and close. Wide binaries are stars in large, long-period orbits. Such systems probably formed by the accidental gravitational capture of two stars born separately. These stars will evolve independently, as two separate single stars. That they are a gravitational pair will not concern us much here. Of greater interest, because of the effect on the evolution of the stars, are the close binaries. These systems probably formed by the fragmentation of an initial single protostellar clump. Triple and quadruple systems probably formed in the same way. These close pairs are of particular significance because the presence of a nearby companion profoundly alters the course of stellar evolution.

2. Stellar Orbits

The force of gravity and the principles of conservation of linear and angular momentum govern the orbits of a pair of stars. Recall from Chapter 1 that linear momentum is the product of mass multiplied by velocity, whereas angular momentum, or spin, is the product of the mass, the velocity, and the size of the object under consideration.

Orbits of stars are very nearly ellipses. This is not exactly true if one considers the small effects of the complete theory of gravity as described by Einstein's general theory of relativity, but the assumption that orbits are ellipses is adequate for all our purposes now. We will mostly consider orbits that are the simplest special case of ellipses, namely circles. Two stars orbit one another on elliptical paths around a common *center of mass*. This center of mass can drift through space, but for simplicity we will pretend that there is no net motion of the two stars. Although the two stars share the same sense of the orbit, for instance clockwise, at any given moment, the individual stars move in opposite directions in their mutual orbital dance. They must do so to conserve the linear momentum, to keep the net momentum constant and equal to zero. If they moved in the same direction, the momentum would be first directed in one direction and later in another in violation of the principle of conservation of momentum. Nature does not allow such behavior. The sizes of the orbits are different if the masses of the two stars are different. Again to balance momentum, the smaller-mass star must move faster in the opposite direction to offset the momentum of the larger-mass star. The *period*, or the time for the stars to complete an orbit, must be the same for both. When the first star has traveled all around the second, the second cannot have traveled only part way around the first. If the smaller star moves faster but takes the same amount of time to complete an orbit as the more massive star, then the smaller star must cover more distance. The orbit of the smaller star must be larger.

Similar laws govern the orbits of the planets around the Sun. The planets move in relatively large orbits about the center of mass that lies between the planets and the Sun. At the same time, the Sun is not completely stationary but moves in a tiny orbit about the center of mass. The size of the Sun's orbit is about the same as the physical size of the Sun itself. The Sun moves at about 30 miles per hour, a small but measurable speed. The presence of large planets around nearby stars was recently established with techniques to measure such speeds. The Sun's orbit is fairly complex in detail. Although the Sun mostly responds to Jupiter, the Sun is trying to orbit around the center of mass of nine planets at once.

Using the data on planetary motion carefully garnered by his mentor, the Danish astronomer Tycho Brahe, Johannes Kepler deduced empirically that planets move on ellipses (his first law) and that the period of the orbit is simply related to the size of the orbit (his third law). The angular momentum of the orbital motion depends on the mass and velocity of the two stars, just as the linear momentum does. The angular momentum also depends on the size of the orbits. For this reason, the angular momentum helps to determine exactly how big the orbits will be for two stars of given mass and velocity. Kepler's second law of orbital motion comes about because the angular momentum of each star about the center of mass is constant.

With the help of Newton's law of gravity, we now interpret Kepler's third law as saying that the square of the period, P, of an orbit is proportional to the cube of the size, a, of the orbit divided by the total mass, M_{tot}, of the two orbiting stars. This law and the understanding of it are crucial in astronomy. The relation between period, orbital size, and mass provides the only reasonably direct way to measure the masses of stars. For two stars in a binary system, astronomers can measure the period fairly easily and the separation between the two stars with some difficulty. These two pieces of information and Kepler's third law as codified by Newton determine the total mass of the system. Astronomers must obtain other information to suggest how much of the mass is in each star. One of the reasons why the study of double-star systems is so important is that double-stars provide direct information on the masses of stars.

3. Roche Lobes: The Cult Symbol

Before reading this section you must assume the posture and repeat the oath of secrecy. Curl your right arm over your head and place the fingers of your right hand on your left shoulder. Then curl your left arm so that the fingers of your left hand also touch your left shoulder. Now whisper loudly, "I solemnly swear not to reveal what I am about to learn to anyone upon penalty of being ridiculed by my peers." As we proceed with this chapter you will find that the significance of the posture is that your brains were about to undergo mass transfer onto your shoulder.

For two stars in a binary system, each reaches out to gravitationally dominate some region beyond its own surface, as shown in Figure 3.1. The more massive star, the star on the left in Figure 3.1, has a larger sphere of influence. If one carefully maps the regions of influence of each star, accounting for the complexities of the fact that each star is moving in orbit, you find that the boundary of the two regions, seen in cross section, resembles a figure eight turned on its side. The two halves of the figure are called *Roche lobes* after the German scientist who first worked out their mathematical form. The physical importance of these gravitational lobes is so great that no lecture on binary stars can continue without a sketch of the famous figure. For this reason one of our colleagues refers to this sketch as the "cult symbol" of the priesthood of the binary-star specialists.

The neck of the figure where the two lobes join is called the first or *inner Lagrangian point,* after the French mathematician Lagrange who also studied these systems. This point represents the position in space where the pull of gravity from the two stars just balances. A slight tip in either direction will send a bit of matter falling toward one star or the other. Beyond the surface of the Roche lobes, matter would belong to neither star but would be comfortable to orbit both of them. On a line extending out through the stars are the second and third Lagrangian points. Beyond these points, centrifugal forces overwhelm the combined pull of gravity of the two stars and tend to throw matter out of the system completely. At right angles to the line

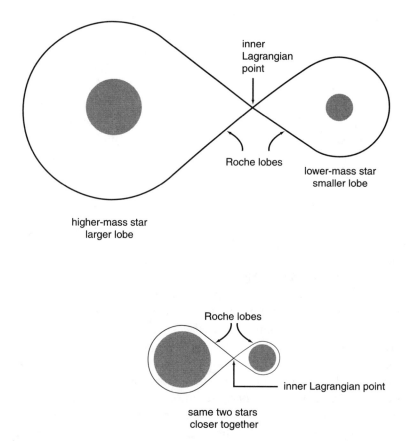

Figure 3.1. The upper diagram shows the Roche lobes, the regions of gravitational domain, around two orbiting stars. The lower diagram shows the same stars in closer orbit. Note that the Roche lobes are always roughly as large as the distance between the stars, but that the star with the larger mass always has the larger gravitational domain and hence the larger lobe. Both of the lobes are smaller if the stars are closer together.

between the stars, one finds the fourth and fifth Lagrangian points. These are of little interest to us in the present context, but these Lagrangian points are potentially important in the subject of space colonization, as past members of the L5 Society will know (the fifth Lagrangian point was their cult symbol). The fourth and fifth points are locations at which a third body is locked in a stable position in the gravity of the two main objects. The idea is that this would be a good place to locate an artificial space colony between Earth and the Moon.

4. The First Stage of Binary Evolution: The Algol Paradox

One of the first lessons learned in the study of binary star systems is that the presence of a companion alters the course of evolution. Recall one of the most important aspects of the evolution of single stars. More massive stars have more fuel to burn, but they burn the fuel at a profligate rate. As a result, massive stars live a much shorter time than smaller-mass stars that hoard their meager allotment of hydrogen fuel. Given this most important lesson, how are we to understand the demon star Algol?

The star Algol presents a blue-white appearance to the eye. Algol also appears to be brighter and then dimmer every few days. When it is dimmer, it appears to be a little redder. In some early cultures a red, winking light in the sky did not bode well. Thus Algol acquired the name the demon star, "Algol" being the Arabic word for demon. We now understand that Algol is a binary system. The red appearance comes because one of the stars is an evolved red giant. The winking derives from the fact that we happen to be looking almost edge-on to the orbits of the stars and hence witness the eclipses as each star in turn moves in front of the other. The slight reddening occurs because one of the stars is a red giant, and we see more of its light and less from that of the blue-white companion when the red giant is in front. We can go a step farther. Because one star has already evolved and has become a red giant, and the other star is still on the main sequence, we know which is the more massive. The red giant has evolved first so the red giant must be the more massive.

Wrong! From the measured period, some astronomical tricks, and Kepler's ever-handy third law, we can work out the masses and find that the red giant has a mass of about 0.5 solar mass, whereas the main sequence star has 2–3 solar masses. This is the *Algol paradox*. How can the evolved star be the less massive one?

To resolve the paradox, we hold firm to the idea that the red giant must originally have been the more massive in order for it to have evolved first. Our basic lessons are impeccable there. The key to resolving the paradox is that, unlike most single stars, close binary stars do not retain the mass with which they were born. When two stars are close together, as in the Algol system, one star can transfer mass to the other. The star that was the most massive became a red giant and then transferred mass to the other star until the mass ratio reversed completely: the originally more massive star became the less massive, and the originally less massive became the more massive.

5. Mass Transfer

To see how this process of mass transfer occurs, we must return to the meaning of the cult symbol, the Roche lobes. Even in a binary system, evolution begins on its normal course. Two stars in a close binary system are presumably born out of the same fragment of interstellar gas, and hence born at the same time. These are fraternal, not identical, twins, however. The chances of the stars being of identical mass are virtually nil. One star will be appreciably more massive than the other. The more massive star uses up its supply of hydrogen in the center and begins to evolve first. The core shrinks, the envelope expands, and the star begins to become a red giant. The more massive star has a greater gravitational domain and hence the larger Roche lobe. The size of the lobe is still finite, however – roughly the same size as the distance between the stars, as you can see from Figure 3.1. As long as the stars are closer together than the eventual size the red giant would normally attain, the presence of the companion star interrupts the normal evolution. This interruption of the evolution is the basic criterion for whether a given binary system is categorized as a close binary system.

The story must change when the more massive star expands to the point where that star fills its Roche lobe. The internal forces of core contraction continue to cause the envelope to expand. As the outer parts of the star pass beyond the Roche lobe, however, they are beyond the gravitational influence of the star from which they came. When that happens, the matter that has moved out beyond the star's Roche lobe no longer belongs to that star. Some of the mass will take up a swirling orbit around both stars, but a great deal will find itself forced through the neck at the inner Lagrangian point joining the Roche lobes of the two stars. Matter that passes through the inner Lagrangian point now finds itself within the gravitational region of influence of the second star. The more massive star transfers matter through the inner Lagrangian point to the other star.

This mass-transfer process is unstable and results in rapid changes in the stars. To see this, recall the nature of the Roche lobes. The more massive star has the larger lobe. The star evolves, fills its lobe, and begins to lose mass. As the star loses mass, the star has a smaller region of influence, so its Roche lobe shrinks, as illustrated in Figure 3.2. Matter otherwise safely attached to the star finds itself cast adrift because the Roche lobe is smaller. That causes the loss of even more mass, resulting in an even smaller Roche lobe. A positive feedback operates in the sense that the more mass the star loses, the more it is forced to lose. The more massive star only approaches the condition of mass loss on the relatively slow time dictated by the contracting of the core. After the mass loss starts, however, it continues at a rapid pace independent of any internal changes in the structure of the star.

This rapid phase continues until the stars have equal mass – the bigger one having lost mass, and the smaller one having gained it. Up to this point, the stars have been spiraling closer together as the star transferred mass. This is due, in large part, to the conservation of angular momentum. Mass is being added to the less massive star that moves with a higher velocity. Higher mass at a higher velocity would

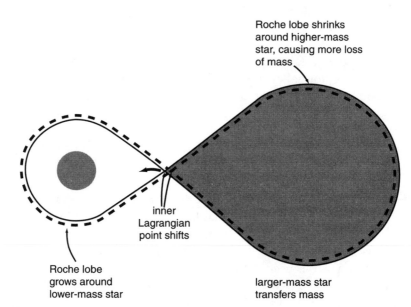

Roche lobe shrinks
around higher-mass
star, causing more loss
of mass

inner
Lagrangian
point shifts

Roche lobe
grows around
lower-mass star

larger-mass star
transfers mass

Figure 3.2. When the more massive star in a binary system loses mass, the process is unstable. As the more massive star loses mass, its Roche lobe becomes smaller, thus biting more deeply into the mass-losing star and causing even more mass loss. This effect is exacerbated because the requirement for angular momentum to be conserved also forces the stars to spiral closer together, making both Roche lobes smaller. As mass is transferred, the location of the inner Lagrangian point shifts to reflect the changing balance of the mass.

mean excess angular momentum. The stars correct this problem by moving together since a smaller-size orbit has less angular momentum. That the stars get closer together during the rapid phase of mass transfer only enhances the rate of transfer because the Roche lobes of both stars, particularly of the star losing mass, get smaller as the stars move together (Figure 3.1).

Although it does slow down, the mass transfer does not halt after the stars attain the condition of equal mass. Now conservation of angular momentum works to make the stars spiral apart. As the star continues to lose mass, it is now the smaller-mass, higher-velocity star. Angular momentum would decrease if the star did not move to a larger-size orbit as mass moved from the more quickly moving star to the slower. The tendency for the stars to move apart once the mass-losing star becomes the less massive means that, as the star loses mass, its Roche lobe gets bigger, not smaller. In order for the mass transfer to continue, the star must expand to fill its new larger Roche lobe. This expansion occurs, but only on the longer time of the internal changes of the structure as the core contracts. The mass transfer no longer involves a positive feedback, and it is thus slower; but mass transfer will continue until the star ceases its attempt to become a red giant. The Algol system is presumably in this slow mass transfer phase.

6. Large Separation

When the two stars are of relatively large separation, but still close enough to qualify as a "close" binary, mass transfer does not begin until the more massive star has become nearly a full-fledged red giant. In this case, the mass-losing star will have a large envelope and a tiny core. The mass transfer continues until virtually the whole envelope vanishes and only the core remains.

If the original star was not too massive (less than about 8 solar masses), the core left behind will be dense and supported by the quantum pressure. It will just cool to become a white dwarf. The result will be a tiny white dwarf orbiting around a more massive main sequence star. The main sequence star will have grown in mass because it is the repository of much of the envelope matter that originally shrouded the white dwarf.

A more massive star (one originally more than about 8 solar

masses) can leave a larger core behind. Such a core will be supported by thermal pressure. It can continue to evolve without the envelope by contracting and heating until new nuclear fuels ignite in its center. The likely outcome for such a core will be to develop an inner iron core that is susceptible to the inevitable collapse. The situation is then similar to that for single stars. The collapse could create an explosion that would leave a neutron star behind. Alternatively, the collapse could be complete, resulting in the formation of a black hole. The result is that we could reasonably envisage the creation of binary systems with a normal star orbiting any of the types of compact stellar remnants we have discussed: white dwarfs, neutron stars, or black holes. We will discuss these cases in Chapters 5, 8 and 10.

7. Small Separation

If the two stars are too close together, the stars evolve in a very different way. Stars swell a bit in size as they consume their hydrogen on the main sequence. This is because the helium that builds up in the center occupies less volume than the hydrogen did. When the helium contracts, the gravitational energy transfers to the outer parts of the star, causing those parts to gain energy, expand, and cool slightly. The process is very similar to that which causes a star to become a red giant, but on a much smaller scale. If the stars are very close together, even this gentle swelling on the main sequence can cause the more massive star to fill its Roche lobe.

The twist comes after the rapid phase of transfer halts, when the two stars have equal masses. Ordinarily, the mass-losing star is a red giant and is evolving internally so rapidly that the mass-receiving star, which is still on the main sequence, is a totally passive partner. In the present case, however, we end up with both stars still on the main sequence. The mass-losing star is evolving slowly, continuing to push mass onto its companion. The evolution of the companion speeds up as it gains mass. Normally, the speed-up is insignificant, but for the case of close stars, the second star also swells to fill its Roche lobe. Each star then tries to transfer mass to the other simultaneously. The situation gets quite messy.

One thing that surely happens with both stars shoving mass beyond their Roche lobes is that material escapes to the region where it surrounds both stars. This matter will orbit in a disk that is in the same plane as the orbit of the two stars and that surrounds both stars.

Matter flows outward into this disk, so such configurations have been dubbed *excretion disks* to distinguish this flow from *accretion disks* where material settles inward. Accretion disks will be the topic of Chapter 4. The system probably ejects some material completely into the surrounding space.

Computer calculations show another interesting possibility. With both stars trying to move mass onto the other, only one can win. The calculations show that the star that had the smaller mass may win this contest, or lose it, depending on your point of view, in the sense that it transfers all its mass to the larger one. The big star consumes the little one! The net outcome is not some exotic binary, but a single star, perhaps surrounded by an excretion disk, the sole evidence of the cannibalism.

8. Evolution of the Second Star

In the standard picture where the star of initially smaller mass remains patiently on the main sequence until the other star completes its evolution, the second star eventually gets its turn. The second star will consume the hydrogen in its center, including perhaps some of that added by the other star. Then the second star will begin to swell as its core contracts, and it, too, will eventually fill its Roche lobe.

At this point, the second star will begin to lose mass to the first. The second star does not particularly care what form its companion is in; it will just proceed to push mass over onto it. From an astronomer's point of view, the results can be quite exciting because the star receiving the mass is a white dwarf, a neutron star, or a black hole – a compact star with a large gravitational field. The effect can be quite spectacular. Astronomers have observed many systems where a star is transferring mass to a compact star. Some of these binary systems with compact stars may have evolved in the rather clean way described in the previous paragraph, with the second star simply swelling to fill its Roche lobe. In other cases, we will see that the actual evolution is probably more complex.

9. Common-Envelope Phase

The principle factor that can spoil the simple picture of one star filling its Roche lobe and transferring matter to the other star that passively accepts the mass is that the second star is unlikely to be a completely passive partner. The mass-gaining star can resist the process, as happened for two main sequence stars very close together. The issue is, if neither star wants the mass, where does it go?

This issue arises more critically for stars that are more compact. For a star of a given mass, whether it is a main sequence star, a white dwarf, or a neutron star, the strength of the gravitational pull depends only on the distance from the center of the star. The gravity does get stronger, the closer one gets to the center of the star. For this reason, the gravity at the surface of a white dwarf is much greater than the gravity at the surface of a normal star of the same mass, and the gravity at the surface of a neutron star of the same mass is greater even yet. The implication is that, if matter falls from a mass-transferring star at a given rate onto a normal star, the impact of the matter with the stellar surface will liberate energy and create luminosity at a certain rate. If the same star transfers mass to a white dwarf at the same rate, the energy liberated when the matter strikes the white dwarf surface will be much greater, thus generating much more heat and a much larger luminosity. The case of a neutron star will be even more extreme. Although a black hole does not have a surface, matter can still respond to the effects of the strong gravity very near the black hole. The result can again be the generation of a large luminosity.

The luminosity generated by the matter that falls in can serve to resist that very infall. The luminosity flooding outward can exert a pressure. In the extreme case, and this case arises in common circumstances for neutron stars, the luminosity can exceed the Eddington limit (described in Chapter 2). This means that the infalling matter is creating a luminosity so great that the resulting pressure is sufficient to prevent the infall! Even in less extreme circumstances, the energy of infall can inhibit the infall. Faced with this resistance, some of the matter will not collect on the mass-gaining star but will go in orbit around both stars.

When this process gets extreme, the matter lost from one star goes predominantly into orbit around both stars, interacts with itself, and bloats to become an approximately spherical (in the imagination of theorists, anyway) bag of gas in which the core of the mass-losing star and the mass-gaining star orbit. The resulting configuration is known as a *common envelope* because the envelope of matter surrounds both stars.

This situation can profoundly affect the orbits of the stars. Now they are not orbiting in the vacuum of space but in a bag of gas. The gas resists their motion, the stars feel friction and drag, and their motion heats the gas. The drag will tend to slow the forward velocity. In the ever-present grip of gravity, the result will be that the stars spiral toward one another and end up orbiting even faster. This will create more friction, heat, and drag and cause the orbits to shrink even faster. The energy and angular momentum lost from the star goes into the common envelope at an ever-increasing rate.

The details of this process are not well understood, but the principle of conservation of energy gives insight into the general nature of the subsequent events. The gravitational energy from the decaying orbits eventually becomes equal to the gravitational energy that binds the common envelope to the two stars. At this point, the energy injected into the envelope by the motion of the stars will be sufficient to blow the envelope away. This process is not an explosion but something more like the ejection of a red-giant envelope to make a planetary nebula. The common envelope will be ejected, and the two stars – whatever configuration they may be in, the core of the mass-losing star and the mass-gaining star – will again orbit in the vacuum of space, but now they will be very close together. Astronomers think this process produces pairs of white dwarfs, neutron stars, and perhaps black holes, in addition to various combinations of these stars and normal stars. We will explore these combinations in Chapters 5, 8, and 10.

10. Gravitational Radiation

Suppose two stars have survived as compact stars, white dwarfs, neutron stars, or black holes that have weathered mass transfer from first the originally more massive star, then the originally less massive star and any intermediate common envelope phase. Now they are orbiting quietly in space. Is this the end of the story? The answer is no!

An important prediction of the general theory of relativity is that gravitational waves spread like ripples through curved space. If a wiggle occurs in the curvature of space, waves will propagate outward carrying off energy and momentum. Imagine an elastic rubber sheet on which you grab a pinch and shake it up and down or the act of poking your finger in the surface of a still pond. Ripples will move outward across the sheet or pond. Ripples in space-time will propagate in the elastic curved space described by general relativity.

Two stars moving in orbit cause a rhythmic change in the curvature of the space around themselves as they circle. The effect is as if you were to twirl a small paddle on the surface of a pond. Ripples spread out across the pond, and gravitational waves spread out through space away from the orbiting stars. The waves carry energy and angular momentum away from the stellar orbits and cause the stars to spiral closer together in the grip of gravity. Eventually, they must collide in some way. In some very special, but important, cases, this loss of energy can determine the life and death of stars. We will discuss these issues further in Chapter 6.

4

Accretion Disks

Flat Stars

1. Background Perspective

One of the major developments of mid-twentieth-century stellar astrophysics was the understanding that there is often a third "object" in a binary star system, especially in a system undergoing mass transfer. Matter from one star swirls around the other forming a configuration known as an *accretion disk*. Such disks were first recognized in the study of white dwarfs in binary systems. With the advent of X-ray astronomy, it became especially clear that accretion disks play a prominent role in binary systems containing neutron stars and black holes. In many circumstances, the accretion disk is the primary source of visible light; in others, the disk is also the primary source of X-ray radiation; and in yet others, the disk channels matter into streams of outgoing material and energy. One dramatic fact is that, without accretion disks, we would not yet have discovered any stellar-mass black holes.

One star in a binary system must undergo mass transfer to feed the disk the matter needed for the disk to exist at all. The disk forms around the star receiving the transferred mass. An accretion disk thus also depends on a more ordinary star (considering black holes to be "ordinary" in this context!) for the gravity to hold the disk together. Given this support from the two stars in the binary system – one to provide matter, one to provide gravity – the accretion disk then effectively has a life of its own. The accretion disk has a structure and evolution that depends only incidentally on the properties of the star at its center or the one providing it mass. The disk is almost like a separate star, a flat star. The disk generates its own heat and light and can have eruptions that have nothing directly to do with either of the stars.

2. How a Disk Forms

In common situations, the matter that feeds the disk flows from the companion star through the inner Lagrangian point that connects the two Roche lobes in the binary system. The structure of the inner Lagrangian point makes it act like a nozzle. The matter thus leaves the mass-losing star in a rather thin stream in the orbital plane of the two stars. In reality, the matter may spray in a messier fashion, but most of the matter remains in the orbital plane. If the two stars were stationary, this matter would flow from one star directly along the line connecting the centers of the two stars and strike the mass-gaining star. In a binary star system, however, the stars are constantly moving in orbit, so the mass-gaining star is a moving target. The matter may leave the mass-losing star headed for the other star, but because the other star moves along in its orbit, the transferred matter cannot fall directly onto the mass-gaining star.

If the mass-gaining star is small in radius, and white dwarfs, neutron stars, and black holes all qualify in this regard, then when the mass flow first starts, the stream of matter will miss the mass-gaining star entirely, passing behind the star as the star moves along in orbit. The gravitational domain of the mass-gaining star captures the matter, however, so the stream circles around and collides with the incoming stream. As this process continues, the flow of self-interacting matter will form first a ring and then a disk. From that point on, the transferred matter will collide with the outer portions of the disk and become incorporated into the disk.

The process by which the self-colliding stream of matter becomes an accretion disk involves the angular momentum of the matter in a crucial way. When the stream of matter first circles around the mass-gaining star, it has a certain angular momentum with respect to the star it orbits. Conservation of angular momentum forces the matter to move in a circular path of a certain size. The size of this path depends on the motion of the two stars. If the matter just stayed in this path, it would form a ring, somewhat like the rings around Saturn. To form a filled-in accretion disk that extends all the way down to the surface of the star, the material must settle to ever smaller orbits. Matter in a smaller orbit will have a higher velocity, but the net effect is still to have a smaller angular momentum. Only if the orbiting matter loses some of its angular momentum can the matter move inward and settle onto the central star. The angular momentum must be conserved in the whole binary system, but the matter in the disk must transfer some of its angular momentum elsewhere. Without this loss of angular momentum by the disk matter, the matter would stay in a ring. With a loss of angular momentum, the matter can settle inward, forming a full-fledged accretion disk.

One of the remarkable things about accretion disks is that they are structured in just such a way to provide for this transfer of angular momentum elsewhere, as illustrated in Figure 4.1. Kepler's third law tells us that because the matter in the disk that is closer to the central star must have a smaller orbit where the gravity is higher, matter in a smaller orbit must move faster. Thus each piece of material in the disk finds the material just beyond moving a little slower, and the material just within its orbit moving a little faster. The result is an inevitable rubbing of all the orbiting streams of material on all the adjacent streams. Each stream is slowed down by the slower, outer, adjacent stream and is thus forced to spiral inward. The result, ironically, is for the matter to end up moving faster because the material picks up energy from the gravity of the central star. This process is fundamentally the same one that caused a star to heat up as it lost energy, as we discussed in Chapter 1. The effect of conservation of energy in the presence of gravity is to gain speed (or temperature) when some energy is taken away from the gravitating matter. The result of the rubbing and slipping inward is that the matter gradually settles onto the surface of the star. This process of gradual addition is known by the general term *accretion*, and hence the resulting flat structures are known as accretion disks. The angular momentum that is lost from the disk is gained by the orbiting stars or perhaps blown from the system by winds. The total angular momentum is, in any case, conserved.

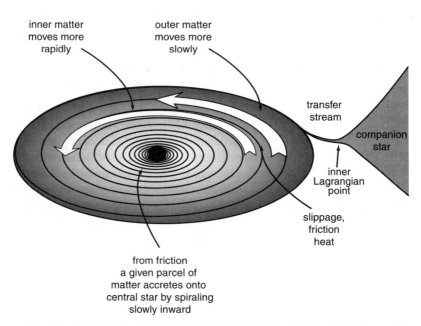

inner matter
moves more
rapidly

outer matter
moves more
slowly

transfer
stream

companion
star

inner
Lagrangian
point

slippage,
friction
heat

from friction
a given parcel of
matter accretes onto
central star by spiraling
slowly inward

Figure 4.1. The orbiting of matter in an accretion disk naturally makes the matter that is farther from the central object move more slowly than matter that is nearer to the center. This creates a constant "rubbing" of the streams of matter. The rubbing results in friction and heating of the matter so that it radiates. The friction also causes the matter to slowly spiral down onto the central object.

3. Let There Be Light – and X-Rays

The other important aspect of the inescapable friction that causes the matter to spiral in and accrete on the star is that friction heats the matter in the process. The heat escapes as radiation that astronomers can study. Because the orbital velocities are lower in the outer portions of the disk, the amount of slipping, friction, and heat are relatively low. The outer portions of the disk are typically about as hot as the surface of the Sun and emit much of their energy in the optical portion of the spectrum. In the middle of the disk, the velocities are higher, the friction and heat are greater, and the energy characteristically emerges in the ultraviolet portion of the spectrum. This is the end of the story if the mass-gaining star is a white dwarf, because the matter spiraling inward in the accretion disk collides with the white dwarf before the matter gains substantially more energy. For neutron stars and black holes, however, conditions can get even more extreme. The velocity of the spiraling matter can get near the speed of light. The frictional heating is immense. The matter gets so hot that the radiation emerges as X-rays, as shown in Figure 4.2. This is one reason that the search for neutron stars and black holes in binary stars requires X-ray instrumentation. Those instruments work best on satellites above the absorbing atmosphere of the Earth, so the astronomy of neutron stars and black holes has been primarily one of the space age. We will tell this story in Chapters 8 and 10.

4. A Source of Friction

The study of these flat stars called accretion disks has been a major undertaking in astronomy over the last three decades. The understanding of accretion disks is still in a somewhat crude state. The situation is analogous to the early days of stellar evolution when there was an understanding of the balance between pressure and gravity, but the power source of stars was not known. The problem was that nuclear physics had not been invented. For accretion disks, the physics

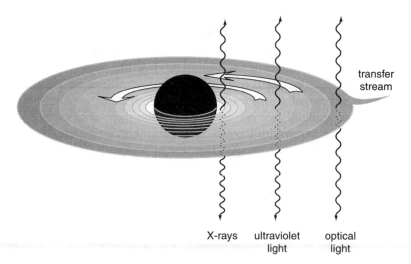

Figure 4.2. Because the orbital velocity of the matter in an accretion disk increases inward, the resulting friction and heat increase, and the resulting temperature of the orbiting matter rises. The outer parts of an accretion disk typically radiate in the optical and the middle parts in the ultraviolet; the innermost parts, if they exist, radiate X-rays.

that determines the heating of the disk is known in principle, but its application is very complex in practice. The net effect is much the same. The drawback for accretion disk theory is that we do not know the nature of the friction, and so the mechanism to generate heat in the disk remains an important unknown.

We know that the normal microscopic rubbing of molecules in a gas is vastly insufficient to provide the friction and heat in observed accretion disks. Rather, the friction must come from large-scale roiling in the disk. Work of the last few years has provided evidence that magnetic fields must play a role in this process to generate the turbulent roiling motions and to couple one eddy in the complicated flow to another to make the interaction and the friction effective.

One compelling theory is that any magnetic field in the disk becomes naturally and unavoidably stretched in the orbiting matter in the disk. A simple analog of the process is to imagine a satellite connected to the space shuttle by a stretchy spring, as shown in Figure 4.3. If the satellite travels in a slightly lower orbit, the satellite will move faster than the shuttle. This will increase the tension in the

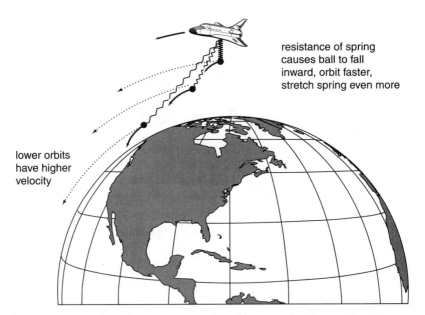

resistance of spring
causes ball to fall
inward, orbit faster,
stretch spring even more

lower orbits
have higher
velocity

Figure 4.3. A satellite in a lower orbit than the space shuttle would orbit more rapidly. If the satellite is coupled to the shuttle by a spring, the spring will add some drag, causing the satellite to settle inward and to orbit even faster. The process will run away until the satellite burns up or the spring breaks.

spring and result in a "drag" on the satellite. Normally if there is drag on a moving object, it will slow down. In the case of an orbiting satellite, however, the drag of the spring that slows the satellite leaves it with too little speed to maintain its orbit. The satellite must settle into a lower orbit where gravity is stronger and things orbit with even higher velocity. The net effect of the drag by the spring is to make the satellite settle into a lower orbit, closer to the Earth, where it moves even faster! This is yet another example of the working of conservation of energy (and angular momentum) when gravity is present. When a gravitating system loses energy, it heats up (like a star) or moves faster (like the satellite). When the satellite settles inward, it gains an even larger relative velocity with respect to the shuttle. The satellite will thus move even farther from the shuttle, increasing the tension in the spring and increasing the drag even more. The process clearly runs away, until the satellite burns up or crashes into the Earth or the spring breaks. In accretion disks, the shuttle and the satellite are represented by two blobs of matter in different orbits, and the spring depicts a line of magnetic force connecting them, as illustrated in Figure 4.4. Any attempt to connect the blobs by means of the magnetic field will cause them to orbit even farther apart and increase the tension in the magnetic field until it snaps. The snapping magnetic field can put energy into the roiling matter and drive the turbulent motions that make the friction and heat.

This magnetic coupling process must exist in accretion disks and play a role in their friction. It may not be the whole story because this theory does not seem to account for the full variability of the friction deduced from observations of accretion disks. Other theories propose that dynamos that generate magnetic fields spontaneously arise in the disk. Energy from the orbiting stars powers the dynamos. Eventual understanding will probably combine both of these ideas and more.

5. A Life Of Its Own

One of the most compelling pieces of evidence that an accretion disk can have its own behavior is when a disk flares with increased brightness. In most systems, the matter flows from the companion star so rapidly that the accretion disk is kept hot and ionized, and the disk radiates steadily. In other systems, however, the flow of matter being transferred is not sufficient to keep the disk in the hot, bright state,

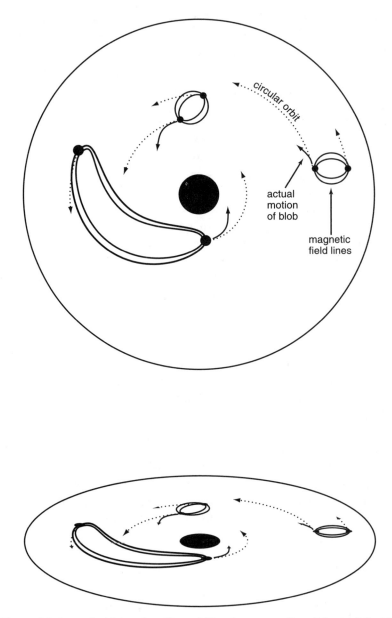

Figure 4.4. Separate blobs of matter orbiting in an accretion disk and linked by magnetic fields behave in a manner that is analogous to the shuttle, satellite, and spring combination shown in Figure 4.3, with the magnetic field playing the role of the spring. The pull of the magnetic field on the inner, more rapidly orbiting blob, will make it settle inward, stretching the magnetic field and causing even more drag on the inner blob and more settling. The stretched magnetic field will eventually "snap," and the energy released will cause the matter to roil, to heat, and to radiate.

and the disk flares only occasionally. Astronomers observe this behavior in disks around white dwarfs, neutron stars, and black holes. There may be a variety of phenomena involved in this flaring, but there is one process that certainly happens in common circumstances. Under certain conditions, the flow of matter in the disk cannot be steady. Rather, the matter stores and then flushes from the disk. The flushing stage is especially bright and causes the flare of radiation. This process is rather independent of the two stars that feed the disk and hold it together. The timing of the flare events and their specific observational features do depend on the central star. If the central star is a white dwarf, astronomers call the flaring a *dwarf nova* (Chapter 5). If the central star is a neutron star or black hole, the flushing of the disk results in an *X-ray transient* (Chapters 8 and 10).

The theory behind this behavior is that the generation of the friction and heating in the disk depends on the temperature in the disk. When the disk is at a low temperature, less than that at the surface of the Sun, the matter in the disk is rather transparent. Any heat generated by the low friction can easily escape as radiation, thus maintaining the low-temperature state. In this low-friction state, there is little tendency for the matter to settle inward, but new matter flows from the companion star. The addition of matter increases the density of the material in the disk. As the density increases, however, the matter becomes more opaque, radiation cannot escape so easily, and the temperature must rise. This leads to a runaway process. The reason is that, as the matter heats, it becomes even more opaque to radiation. This traps more heat, leading to a greater opacity and an even greater trapping of the heat.

The result is that the disk can exist in a cool, barely accreting state, with low luminosity, until enough density accumulates to trigger this heating runaway. The beginning of such an outburst is illustrated in the top two panels of Figure 4.5. A wave of heating runs through the disk. The wave can begin on the outside of the disk, as shown in the second panel of Figure 4.5, or deeper down in the disk, depending on circumstances. The disk suddenly becomes very hot and very bright. The disk reaches maximum brightness when the heating fully envelopes the disk as shown in the middle panel of Figure 4.5. The friction increases dramatically in the hot state, and so material that had accumulated in the outer parts of the disk rapidly moves inward. Ironically, this motion of the matter in the disk shuts the process off. As the outer portions of the disk thin out, they become more transparent

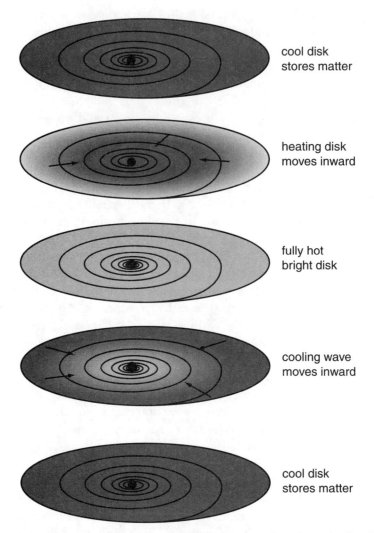

cool disk
stores matter

heating disk
moves inward

fully hot
bright disk

cooling wave
moves inward

cool disk
stores matter

Figure 4.5. When mass is fed into an accretion disk at a rather slow rate, the disk goes through a cycle of cool, dim, storage and hot, bright flushing of matter. (top) Most of the time the disk matter is cool and rather transparent, so little of the matter added from the companion star flows through the disk. (second) As matter accumulates, the density rises, and the disk turns more opaque, trapping the heat. This leads to a heating instability and a heating wave that propagates through the disk. (middle) When the disk is fully heated, it is temporarily very hot and bright, the peak of the flare. (fourth) As matter settles inward, the outer parts thin out, turn more transparent, and cool. A cooling wave moves inward through the disk. (bottom) After the whole disk has cooled, the storage process begins again.

again. They can radiate more easily, lose their heat, and lower the temperature. Now the inverse process sets in. As the temperature drops, the material becomes less opaque and more transparent, and this leads to a greater loss of heat, lower temperature, more transparency, and even greater loss of heat. A wave of cooling sets in from the outer parts of the disk that thin out first. This is illustrated in the fourth panel of Figure 4.5. The cool front sweeps inward, causing the majority of the matter in the disk to settle back into the cool storage state, as shown in the last panel of Figure 4.5. After an interval of storage, the cycle will then repeat.

The net effect is that the disk can exist in its cool storage state for a considerable time. The amount of time depends on circumstances, but the interval can vary from weeks to decades. The disk may be essentially undetectable during this phase. Then the eruption occurs, and the disk becomes very hot and bright for a short time, typically one-tenth the time the disk was dim, and is readily visible to astronomers. No sooner has the eruption occurred, however, than the disk starts to cool. Astronomers who want to study this transient bright phase must scramble!

An important aspect of this cycle of quiescence and eruption is that the process can be quite independent of the stars in the system. During the whole process, the mass-losing star can be pumping matter in at a perfectly constant rate. The star around which the accretion disk swirls provides a constant gravity. The flaring activity is a feature of the disk alone. In more complex systems, the mass can flow from the mass-losing star at a variable rate. The mass-gaining star can have a hard surface or strong magnetic field of its own (in the case of either neutron stars or white dwarfs). Either of these situations can lead to interesting variations.

6. Not So Flat, Buddy!

Another important idea has emerged in the last few years. The inner parts of accretion disks may not be so flat. Under certain circumstances, as the disk cools after its heating episode, the density can get so low that interactions among the particles are rare, and the efficiency of radiation can drop. This again leads to a retention of heat. The excess heat leads to pressure that causes the disk to swell up and become fatter, as shown in Figure 4.6. If this happens, this portion of

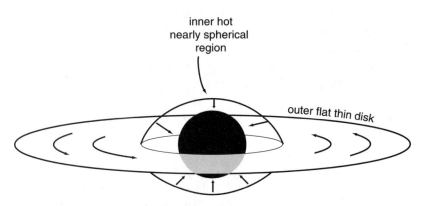

Figure 4.6. The inner portions of accretion disks, especially those surrounding black holes, can retain their heat and swell to become a fat, nearly spherical region. In the outer, thin disk, the matter orbits in a single plane, but in the inner, fat portion, the matter flows nearly radially inward. This radial inflow sweeps heat down into the black hole before it can be radiated away, so the inner regions are relatively dim.

the disk can become so hot that matter and antimatter, electrons and positrons, are created. The disk assumes a more nearly spherical configuration, and matter falls inward on the central star almost uniformly from all directions.

Under these circumstances, the matter can fall in so rapidly that the flow of matter carries the heat generated into the central star before the energy radiates away. This is especially true if the central star is a black hole. The heat energy disappears into the black hole just as the matter itself does. The result is that this fat, inner portion of an accretion disk is especially dim. What little energy leaks out corresponds to especially high energy radiation – high-energy X-rays and gamma rays. There is growing evidence that such regions do form in the centers of accretion disks as they settle back into their storage state. We will discuss this evidence in Chapter 10. One of the outstanding issues, the subject of current research, is when, why, and how a disk makes the transition from the relatively cool, flat configuration of a standard accretion disk to the very hot, fat configuration. Understanding this transition may give new clues for how to find and study black holes.

5

White Dwarfs

Quantum Dots

1. Single White Dwarfs

White dwarfs are certainly the most common stellar "corpses" in the Galaxy. There may be more white dwarfs than all the other stars combined. The reason is that low-mass stars are born more frequently, and low-mass stars create white dwarfs. In addition, after a white dwarf forms, it sticks around, slowly cooling off, supported by the quantum pressure of its electrons. This means that the vast majority of the white dwarfs ever created in the Galaxy are still there. The exceptions are a few that explode or collapse because of the presence of a binary companion. There are probably ten billion and maybe a hundred billion white dwarfs in the Galaxy. Most white dwarfs have a mass very nearly 0.6 times the mass of the Sun. A few have smaller mass, and a few have larger mass. Exactly why the distribution of the masses is this way is not totally understood.

White dwarfs provide clues to the evolution of the stars that gave them birth. To fully reveal the story, astronomers need to probe the insides of the white dwarf. Ed Nather and Don Winget at the University of Texas invented a very effective technique to do this. The technique uses the seismology of the white dwarfs to reveal their interior structure, just as geologists use earthquakes to probe the inner Earth. Under special circumstances, depending on their temperature, white dwarfs naturally oscillate in response to the flow of radiation from their insides. The oscillations cause small variations in the light output. To do white dwarf seismology, careful observations must be made over extended times, days to weeks. The problem is that the Sun rises every day, and that makes observations difficult. Nather and Winget thus invented the "Whole Earth Telescope," in which a network of small telescopes in various sites around the world is coordinated by telephone and the World Wide Web. The trick is that as the Sun rises and the target white dwarf sets in one part of the world, the Sun is setting on the opposite side of the world, and the target white dwarf is rising. With careful planning, the white dwarf can be observed constantly from somewhere on the globe for weeks at a time.

The results have been striking. The Whole Earth Telescope has measured the masses of some white dwarfs with exquisite accuracy. The team has measured the rotation of some of the stars and probed the inner layers of carbon and oxygen. The outer layers, thin shells of hydrogen and helium, have provided clues to the birth of the white dwarfs. By these techniques and others, measurements of the ages of some of the white dwarfs are possible.

Measuring the ages of the white dwarfs is especially interesting because the ages reveal the history of the Galaxy. Because essentially all the white dwarfs ever born are still around, the white dwarfs can tell the story of when the first white dwarfs formed when the Galaxy itself was young. The white dwarfs cool steadily, but they cool slowly. The oldest, coolest white dwarfs are dim and difficult, but not impossible, to see. Studies of the oldest white dwarfs reveal that the first white dwarfs formed about 10 billion years ago. The Galaxy itself presumably formed only a few billion years before that. This argument leads to the conlusion that the Galaxy is relatively young compared to some estimates. The exact age of the Galaxy remains uncertain, but estimating its age with white dwarfs is now an established method.

2. Cataclysmic Variables

A significant number of the white dwarfs in the Galaxy are not alone, but in binary systems. These white dwarfs are especially interesting in the context of this book because they share properties with more exotic objects like neutron stars and black holes in binary systems.

Most of the white dwarf binaries are the result of the first stage of mass transfer when the originally most massive star forms a white dwarf core and transfers the remainder of its mass to a stellar companion. In some cases, both stars have undergone mass transfer leaving two white dwarfs in orbit.

Some of the most common and interesting examples of the second stage of mass transfer are the *cataclysmic variables*. These variable "stars" are all binary systems in which mass flows from one star, first into an accretion disk and then onto a white dwarf. The basic components of a cataclysmic variable system are illustrated in Figure 5.1. The star losing the mass is often a small main sequence star that sometimes has less mass than the companion white dwarf. Emitted radiation tracks the stream of material passing through the inner Lagrangian point and merging with the disk. Most of the light from a cataclysmic variable comes from neither the white dwarf nor the mass-losing star but from the so-called *hot spot* where the transfer stream collides with the outer edge of the accretion disk. This collision is very energetic and so produces a great deal of heat and light. Some light also comes from the friction and heating in the inner reaches of the accretion disk itself as described in Chapter 4.

Several types of cataclysmic variables exist. The types are differentiated by their specific observational properties and the mechanisms thought to cause their variability. Cataclysmic variables all fall under the general category of the novae, or new stars. This is because historically the brightest flares would cause a "new" star to appear where none had been seen before. The star system is not new, of course,

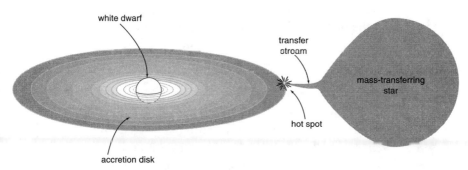

Figure 5.1. Schematic illustration of a cataclysmic variable. The basic components are a star that fills its Roche lobe and transfers mass through an accretion stream, the bright hot spot where the stream strikes the outer rim of the accretion disk, an accretion disk, and a central white dwarf.

merely below the threshold of detectability until the system flares. The phrase "supernova" is an offshoot. For a long time all suddenly flaring events that caused a new star to appear were classified with the same general term, "nova." With the discovery that some events were in distant galaxies, and hence intrinsically very much brighter, shining over great distances, the term "supernova" was applied. We now know that novae and supernovae involve very different phenomena, although they are not completely unrelated. Novae might eventually turn into supernovae, and some novae involve thermonuclear explosions.

Dwarf novae are the most gentle of the cataclysmic variables. Dwarf novae flare up irregularly to be about ten times brighter than they usually are. The flares occur with intervals of weeks to months and last for days to weeks at a time. This interval is too short to build up any reservoir of thermonuclear fuel. The energy involved comes from heating as material from the mass-losing star settles in the gravitational field of the white dwarf. There are two competing ideas of how the flare occurs. One is that the mass-losing star undergoes surges that throw over extra mass from time to time. The problem with this picture is that one would expect the hot spot to flare first, before the disk, but this is not observed. The alternative is that matter piles up in the accretion disk until some instability causes the matter to suddenly spiral down toward the white dwarf leading to an increase in the frictional heating and the light output in the process.

Detailed studies suggest that the disk-heating instability described in Chapter 4 (see Figure 4.5) is the primary cause of dwarf novae. Matter piles up in the disk in a cool, dim storage phase until the disk becomes opaque and traps the heat. This very heating causes an increase in the opacity, yielding more heating, more friction, and yet more opacity. The result is a rapid transition of the disk to a hot bright state. When the central star is a white dwarf, the observed result is a dwarf nova outburst. During the outburst, the extra luminosity will heat the surface of the companion star and may cause the companion to transfer more mass. Both suggested mechanisms may thus play some role in the dwarf nova outburst mechanism.

Recurrent novae flare to become about a thousand times brighter than the conditions prior to the outburst. These flares occur every 10–100 years. The mechanism of the outburst is unknown. Although both kinds of systems involve mass transfer through an accretion disk onto a white dwarf, dwarf novae do not have recurrent nova outbursts, nor vice versa. The difference may follow from the rate of mass transfer. If the rate is fast enough, the disk will steadily channel all the mass to the white dwarf. The disk will not have the luxury of waiting until enough matter has collected to begin to drop the matter onto the white dwarf. A faster mass transfer rate might explain why a recurrent nova does not undergo dwarf nova outbursts, but that does not explain the nature of the recurrent nova outbursts.

Classical novae, or in casual terms, novae, flare from ten thousand to a hundred thousand times brighter than their normal state. None has ever been seen to recur. The suspicion that classical novae repeat on intervals of about 10,000 years has been around for decades. There is, however, little direct evidence for that particular time scale, which is too long for the brief recorded history of astronomy. The established evidence, both observational and theoretical, is that the mechanism of the classical nova outburst is a thermonuclear explosion. The idea is that as matter flows from the companion star, the matter settles onto the white dwarf in a dense layer supported, as is the white dwarf, by the quantum pressure, as shown in Figure 5.2. The inner white dwarf is probably composed of carbon and oxygen that require extreme conditions to ignite and burn. The material collecting on the outside is hydrogen, which burns more easily. As the hydrogen collects, the density and temperature increase until the hydrogen ignites. Because the hydrogen is supported by the quantum pressure, the thermonuclear burning does not increase the pressure and hence cannot at first cause expansion and cooling. Rather, the burning is unregulated, and an explosion ensues. The explosion does not involve the whole star like a supernova, only the outer layers. Nevertheless, the result is spectacular, giving a great flare of light and blowing matter off the surface of the white dwarf at high velocities. If the current theories are correct, the white dwarf will then begin to accumulate more hydrogen from its obliging companion until the conditions are yet again ripe for an explosion.

3. The Origin of Cataclysmic Variables

"Careful readers" (to which class the author never belonged) may have noticed that they were sandbagged earlier in the first general description of cataclysmic variables. The sleeper was the comment that in most cataclysmic variables the star losing mass is a small main

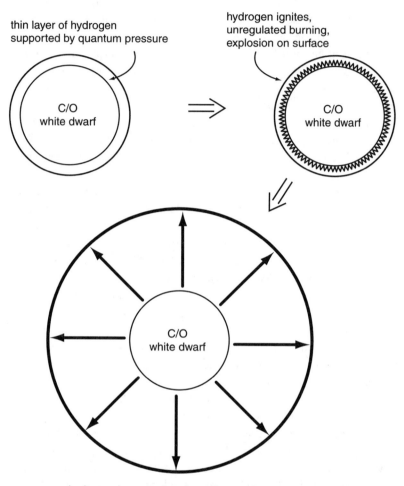

thin layer of hydrogen
supported by quantum pressure

hydrogen ignites,
unregulated burning,
explosion on surface

C/O
white dwarf

C/O
white dwarf

C/O
white dwarf

hydrogen layer, and some carbon and oxygen,
blown into space

Figure 5.2. The mechanism of a classical nova explosion. (top left) Hydrogen from
the companion star passes through the accretion disk and accumulates in a thin,
quantum–pressure supported layer on the surface of a white dwarf, often composed
of carbon and oxygen. (top right) When the density and temperature in the hydrogen
layer get large enough, the hydrogen will begin thermonuclear burning. Because the
hydrogen is supported by the quantum pressure, there will at first be no mechanical
response, the shell will just get hotter, and the burning will be unregulated. This will
result in an explosion. (bottom) The explosion will blow the hydrogen layer into space
along with some of the carbon and oxygen from the central white dwarf.

sequence star. Let us think that through. If a small main sequence star is losing mass, the star must be filling its Roche lobe. Because the star is not large, the lobe must be small, which means that the stars – the main sequence star and the white dwarf – must be very close together, almost touching. How then did the white dwarf form in the first place? The separation must have been large so that the progenitor of the white dwarf could form a well-developed core-envelope structure and become a red giant before mass transfer began. If the stars were very close together originally, the big star would eat the little one (Chapter 3, Section 7). No cataclysmic variable system could evolve. The conclusion is that the two stars must have been far apart initially, even though they are very close together now.

What is necessary to perform this bit of stellar legerdemain is to find a way to drag the stars together. The mechanism proposed to accomplish this is the *common envelope*, described in Chapter 3. We have discussed that matter can spill outward to orbit around both stars in a binary system. There is a strong suspicion that when a red giant goes into the first stage of rapid mass transfer, mass flows at such a rate that the second star is glutted. The matter falling on the star causes heat and extra radiation, and the pressure of that radiation will prevent the rapid flow of matter onto the star.

With a red giant pouring forth mass in copious amounts and the companion refusing to accept it, the matter will enshroud both stars. Unlike the case of an excretion disk where the matter orbits both stars in the orbital plane, this great amount of matter will form an approx-imately spherical red-giant-like envelope around both stars. Both the tiny white dwarf core of the original red giant and the innocent main sequence companion will orbit around inside this envelope. The result is a common-envelope or "double-core" system. The main sequence star and the white dwarf are not orbiting in the vacuum of space now but in the frictional medium of their common gaseous shroud. The friction causes the two stars to spiral together.

The developments that follow then are particularly unclear, but speculation goes as follows. The white dwarf and the main sequence star finally get very close together, so close that the Roche lobe of the main sequence star gets smaller than the star. Notice that the star does not evolve and expand to fill the lobe; the lobe shrinks along with the orbit to fit the star. At this point (perhaps from the heat of theoretical astrophysicists waving their arms), a burst of energy blows away the common envelope. As the ejected matter floats away, a fully formed cataclysmic variable emerges. In this view, the system is "born" within the common envelope as a main sequence star already filling its Roche lobe and transferring mass to a white dwarf. The beginning of the transfer of mass from the main sequence star to the white dwarf may be the energy source that ejects the common envelope.

The simplest, cleanest, mass-transfer process to imagine is that the red-giant envelope flows from one star to the other and thus bares the white-dwarf core. The second star subsequently expands to fill its Roche lobe and transfers mass back to the white dwarf to form a cataclysmic variable. This simple picture is probably relatively rare in practice. Even though many details must yet be understood, the formation of most cataclysmic variables probably involves the more complicated common-envelope process.

4. The Final Evolution of Cataclysmic Variables

The ultimate fate of cataclysmic variables is very uncertain. There are two general possibilities. These systems could just fizzle out. The ordinary star could eject its envelope and leave behind a second white dwarf so that mass transfer stopped. Alternatively, cataclysmic variables could end in a cataclysmic implosion or explosion. Even the fizzle could be interesting, involving some fascinating contortions. Let us examine the catastrophic possibilities first, then return to the fizzle.

In some observed cataclysmic variables, the mass of the white dwarf is within about 10 percent of the Chandrasekhar limit, and the mass is increasing steadily. This situation immediately invokes speculation concerning the outcome if the white dwarf reaches the limiting mass. One possibility is that the nuclear fuel of which the white dwarf is composed – for instance, carbon and oxygen – ignites. For a white dwarf near the Chandrasekhar mass limit, the density is very high. With these conditions, the quantum energy of the carbon nuclei can trigger nuclear reactions even if the temperature and the thermal energy are at absolute zero. As we have described many times now, nuclear ignition under conditions where the star is supported by the quantum pressure is very unstable. Ignition of carbon under these conditions would lead to a violent explosion. This explosion would occur in a star devoid of hydrogen, save perhaps for a negligibly thin layer on the surface. Such a picture is the most probable origin of one kind of supernova, as we will explore in Chapter 6.

The white dwarf could possibly be made of iron that disintegrates upon compression, or, more likely, of oxygen, neon, and magnesium, elements that can absorb electrons rapidly. In these circumstances, when the Chandrasekhar limit is approached, the white dwarf may collapse rather than explode. This process will leave a neutron star in orbit around the main sequence star. This collapse may result in the ejection of little or no mass. The energy of the collapse might come out almost entirely in the form of neutrinos, so that there would be little or no optical display. A process this violent, however, is likely to be bright as well.

All these potential catastrophes depend on the mass getting very close to the limiting value of the Chandrasekhar mass, within a percent or so. One interesting open question is whether the mass ever gets that high. Nova explosions certainly blow off matter that has accumulated on the surface of the white dwarf. If all the matter that has accumulated is ejected in the outburst, the mass of the white dwarf will not increase. The situation could be even worse. Nova explosions are observed to expel an excess of carbon and other heavy elements. This strongly suggests that a nova explosion expels not just the outer layer of accumulated hydrogen but also some of the guts of the white dwarf itself. This would mean that the mass of the white dwarf shrinks as a result of nova explosions. If this is the case, the white dwarf will remain in the binary system until the companion star evolves and forms a white dwarf of its own. Thus nova explosions might lead to circumstances where the final fate of the cataclysmic variable is a fizzle rather than a catastrophe.

The cataclysmic variable might fizzle, but the story is not over just because the system produces two white dwarfs. We need to inquire about the ultimate fate of two orbiting white dwarfs. They can no longer evolve on their own. Supported by the quantum pressure, they will just cool off if left to their own devices. The white dwarfs do not remain unmolested, however. As they revolve about, their orbital motion generates gravitational waves. The gravitational waves carry off energy and angular momentum, and the orbit must shrink. As described in Chapter 3, gravitational radiation affects all stellar orbits. Gravitational radiation is a very small effect, so that any other normal interaction between the stars is more important. Only when the two white dwarfs reach a state of total quiescence can the small effect of gravitational radiation become important. This will inevitably happen to two white dwarfs, however, and they must spiral together. The outcome depends on the specific properties of the white dwarfs as stars supported by the quantum pressure.

For a normal star supported by thermal pressure, the addition of mass causes the star to attain a larger radius. Remove mass, and the star shrinks. For a white dwarf supported by the quantum pressure, the opposite situation holds. The addition of mass causes an overall compaction of the star. The star thus shrinks in radius as the mass increases. Removal of mass from a white dwarf allows the star to expand in the smaller gravity and attain a larger radius. This behavior has crucial implications for the ultimate fate of one of the white dwarfs.

The two white dwarfs will spiral together until the separation, and hence the Roche lobes become small enough so that one of the white dwarfs fills its lobe. Which one will that be? The one with the smaller mass has a smaller Roche lobe but a larger radius. The smaller-mass white dwarf will fill its lobe first and begin to lose mass to the larger-mass dwarf. This is not good news for people rooting for the under-dog! As the smaller-mass dwarf loses mass, its Roche lobe shrinks, but its radius gets even larger! The white dwarf will lose mass at an ever more rapid pace. The only outcome can be the disappearance of the small-mass dwarf. The larger-mass dwarf will simply gobble up the smaller-mass one. Some mass may slop out into space, but this will be little consolation to the disappearing dwarf.

The smaller-mass star may not disappear entirely. When the mass of the object gets down to the size of a planet – less than a thousandth the mass of the Sun – its structure may rearrange. If the material becomes rocklike, like the Earth, then the remains of the little white dwarf may cease expanding. The result could be one white dwarf orbited by a desolate rocky chunk. Given sufficient time for gravitational radiation to act, even that chunk could spiral down to the surface of the remaining white dwarf and be consumed.

Alternatively, the process of disrupting the smaller-mass white dwarf may not end gently at all. As the larger-mass white dwarf consumes the smaller-mass one, the larger-mass white dwarf gets more mass, shrinks to a smaller volume, and hence develops a higher density. This increase in density could result in the ignition of carbon burning in the more massive white dwarf. The resulting catastrophic burning in the more massive white dwarf would blow the star apart. This is yet another proposed mechanism to create a certain type of supernova from a white dwarf. We will see this in more detail in Chapter 6.

White dwarfs may just be small quantum-pressure supported dots, but they can do very interesting things. They may hold the key to understanding the fate of the Universe. We will see that in Chapter 11.

6

Supernovae

Stellar Catastrophes

1. Observations

Which stars explode? Which collapse? Which outwit the villain gravity and settle down to a quiet old age as a white dwarf? Astrophysicists are beginning to block out answers to these questions. We know that a quiet death eludes some stars. Astronomers observe some stars exploding as supernovae, a sudden brightening by which a single star becomes as bright as an entire galaxy. Estimates of the energy involved in such a process reveal that a major portion of the star, if not the entire star, must be blown to smithereens.

Historical records, particularly the careful data recorded by the Chinese, show that seven or eight supernovae have exploded over the last 2000 years in our portion of the Galaxy. The supernova of 1006 was the brightest ever recorded. One could read by this supernova at night. Astronomers throughout the Middle and Far East observed this event.

The supernova of 1054 is by far the most famous, although this event is clearly not the only so-called "Chinese guest star." This explosion produced the rapidly expanding shell of gas that modern astronomers identify as the Crab nebula. The supernova of 1054 was apparently recorded first by the Japanese and was also clearly mentioned by the Koreans, although the Chinese have the most careful records. There is a strong suspicion that Native Americans recorded the event in rock paintings and perhaps on pottery. An entertaining mystery surrounds the question of why there is no mention of the event in European history. One line of thought is that the church had such a grip on people in the Middle Ages that no one having seen the supernova would have dared voice a difference with the dogma of the immutability of the heavens. One historian, the wife of one of my colleagues, has an interesting alternative viewpoint. She argues that the people who made careful records of goings-on in medieval Europe were the monks in scattered monasteries. Some of these monks were renowned for their drunken revelries and orgies in total disregard for their official vows of abstinence and celibacy. Would such people have shied from making mention of a bright light in the sky when they kept otherwise excellent records? (Never put it in writing?) The truth may be more mundane, having to do with weather or mountains blocking the view. A report of a few years ago called attention to a reputed light in the sky at the time of the appointment of Pope Leo, but this has not been widely accepted. In any case, there is no confirmed record of the supernova of 1054 in European history.

Five hundred years later, the Europeans made up for lost time. The supernova of 1572 was observed by the most famous astronomer of the time, the Danish nobleman Tycho Brahe. Tycho made the careful measurements of planetary motions that allowed his student, Johannes Kepler, to deduce his famous laws of planetary motion. Tycho also carefully recorded the supernova of 1572. His data on the rate at which the supernova brightened and then dimmed in comparison to other stars gives a strong indication of the kind of explosion that occurred. The heavens favored Kepler in his turn with the explosion of a supernova in 1604. Kepler also took careful data by which we deduce that he witnessed the same kind of explosion as his master. Although there are counterarguments and some controversy, both Tycho's and Kepler's supernovae are widely regarded to be the kind of event modern astronomers label Type Ia.

Shortly after Kepler came Galileo and his telescope and then Newton with his new understanding of the laws of mechanics and gravity. This epoch represented the birth of modern astronomy. Astronomers now have large telescopes, the ability to observe in wavelengths from the radio to gamma rays, and the keen desire to study a supernova close up. Ironically, however, Kepler's was the last supernova to be observed in our Galaxy. Supernovae go off rarely and at random, so a long interval with none is not particularly surprising, just disappointing. We do observe a young expanding gaseous remnant of an exploded star, a powerful emitter of radio radiation known as Cassiopeia A. From the present size and rate of expansion of the remnant, we deduce that the explosion that gave rise to Cas A occurred in about 1680. By rights, this should have been Newton's supernova, but no bright optical outburst was seen. Evidently, this explosion was underluminous. There are reports that Cas A was seen faintly by John Flamsteed, who was appointed the first astronomer royal of England by King Charles II in 1675, but there are questions concerning whether or not that sighting was in the same position as the remnant observed today. Astrophysicists have calculated that supernovae are brighter if they explode within large red-giant envelopes. The suspicion is that the star that exploded in about 1680 may have ejected a major portion of its envelope before exploding or that the star was otherwise relatively small and compact. That condition, in turn, may have prevented Cas A from reaching the peak brightness characteristic of most supernovae. We will see in Chapter 7 that supernova 1987A, the best studied supernova of all time, had this property of being intrinsically dimmer than usual.

All supernovae directly observed since 1604, and hence all supernovae seen by modern astronomers, have been in other galaxies. Any single galaxy hosts a supernova only rarely. Supernovae occur roughly once per 100 years for spiral galaxies like the Milky Way. Astronomers do, however, observe a huge number of galaxies at great distances. The chance that some of these galaxies will have supernovae go off in them is appreciable. Before supernova 1987A, about thirty supernovae were recorded every year. Closer attention was paid to discovering supernovae after supernova 1987A, and the current rate of discovery is about 100 per year. Many of these supernovae are so distant and so faint that scant useful data are obtained from them, but special programs have yielded good data on very distant supernovae. This will be discussed in Chapter 11.

From the studies of supernovae in other galaxies, astronomers have come to recognize that there are two basic types called, cleverly enough, Type I and Type II. This differentiation was first made in the 1930s when Fritz Zwicky began systematic searches for supernovae at Caltech. The categories of supernovae are traditionally defined by the *spectrum* that reveals the composition of the ejected matter. Complementary information is obtained from the *light curve*, the pattern of rapid brightening and slower dimming followed by each event. As more supernovae have been discovered, the dividing lines of this taxonomy have been blurred by events that share some properties of Type I and some of Type II. As for any developing science, one begins with categories and then seeks to replace mere categories with a solid base of physical understanding.

The spectra of Type I supernovae are peculiar in that they reveal no detectable hydrogen, the most common element in the Universe. Some Type I supernovae, called Type Ia, appear in all kinds of galaxies – elliptical, spiral, and irregular. Type Ia tend to avoid the arms of spiral galaxies. Because the spiral arms are the site of new star formation, the suggestion is that Type Ia supernovae explode in older, longer-lived stars. This implies that the progenitor stars of Type Ia supernovae are not particularly massive because massive stars live only a short time. Just how low the mass of these Type Ia supernovae may be is a question of current debate. The light curve for Type Ia supernovae is very identifiable. There is an initial rise to a peak that takes about 2 weeks and then a long slower period of gradual decay over time scales of months that is very similar for all these events. The data recorded by Tycho and Kepler suggest that they both witnessed Type Ia supernovae. No other galactic supernova has sufficient records to make an identification by type. For decades, all Type Ia supernovae were thought to be virtually identical, but more recent careful observations have revealed small, but real, variations among them.

Near the peak of their light output, Type II supernovae show normal abundances in their ejected material, including a normal complement of hydrogen. The material observed at this phase is very similar to the outer layers of the Sun. These supernovae have never appeared in elliptical galaxies. Type II supernovae occur occasionally in irregular galaxies, but mostly in spiral galaxies and then within the confines of the spiral arms. The reasonable interpretation is that the stars that make Type II supernovae are born within the spiral arms and live an insufficient time to wander from the site of their birth. Because they are short-lived, the stars that make Type II supernovae must be massive. The light curve of a typical Type II supernova shows a rise to peak brightness in a week or two and then a period of a month or two when the light output is nearly constant. After this time, the luminosity will drop suddenly and then less rapidly with a time scale of months. This pattern of light emission with time is consistent with an explosion in the core of a star with a massive, extended red-giant envelope, as will be explained in Section 6.

To confuse the issue, one and maybe two other varieties of hydrogen-deficient supernovae were identified in the 1980s. These are called, with a further flight of imagination, Type Ib and Type Ic. The two types are probably closely related. Unlike Type Ia, but like Type II, Types Ib and Ic only seem to explode in the arms of spiral galaxies. Therefore, Types Ib and Ic are also associated with massive stars. Type Ib show evidence for helium in the spectrum near maximum light. Type Ic show little or no such evidence for helium. On the other hand, both types show evidence for oxygen, magnesium, and calcium at later times. This is the strongest argument that Types Ib and Ic are closely related. They show little or no evidence for the strong line of silicon that is a major characteristic of the spectra of Type Ia. Type Ia supernovae show essentially only iron at later times, another factor emphasizing their difference from Types Ib and Ic. The composition revealed by Types Ib and Ic is similar to that expected in the core of a massive star that has been stripped of its hydrogen. In the case of Type Ic, most of the helium is gone as well. This suggests an origin in a star much like a Wolf-Rayet star, but a direct connection to this class of stars has not yet been established. The light curves of Types Ib and Ic are somewhat similar to those of Type Ia, but are dimmer at maximum light.

A bright supernova observed in 1993, SN 1993J, gave yet more clues to the diversity of processes that lead to exploding stars. SN 1993J revealed hydrogen in its spectrum, so this event was a variety of Type II. As the explosion proceeded, however, the strength of the hydrogen features diminished, and strong evidence for helium emerged. In this phase, SN 1993J looked much like a Type Ib. There were a few events like this known before, and several have been seen since. Apparently this star had most, but not all, of its hydrogen envelope removed, probably in a binary mass transfer process. In other cases, the removal of hydrogen is more nearly complete, and in yet others, for the Type Ic, the helium is removed, too. There is yet no direct observational proof for binary companions in Types Ib or Ic or the transition events like SN 1993J, but computer models suggest this is the case for SN 1993J, at least. Strong winds from massive stars could play a role for the Types Ib and Ic, and the relative importance of winds versus binary mass transfer has not been re-solved.

2. The Fate of Massive Stars

The evidence indicates that Types II and Ib/c supernovae represent the explosion of massive stars. These stars have presumably evolved from the main sequence to red giants and have had a series of nuclear-burning stages producing ever heavier elements in the core. Just which massive stars participate in this process is still debated.

One way to deduce the masses of the stars that make supernovae is to examine the rate at which the events occur in various galaxies. The death rate can then be compared to the rate at which stars are born with various masses. We know that there are many low-mass stars born every year in a galaxy like ours and rather few massive stars (*why* this should be true is a question under active investigation). If we consider stars with mass in excess of about 20 solar masses, we find such stars are born, and hence die, too infrequently to account for the rate at which Type II supernovae explode. If we consider stars with less than about 8 solar masses, we find that such stars die in excess profusion. Stars with mass between about 8 and about 20 solar masses are born and die at the rate of about once per 100 years in our Galaxy. This is also the approximate rate at which we deduce Type II supernovae occur. Type II supernovae probably come from stars of this mass range. Many of these stars, particularly on the upper end of this mass range, are thought to form iron cores that collapse to form neutron stars. There is thus a strong suspicion that Type II supernovae leave neutron stars as compact remnants of the explosion, and that the gravitational energy liberated in forming the neutron star is the driving force of the explosion.

The rate of explosion of Types Ib and Ic supernovae is not well known because relatively few of them have been discovered. Their rate is roughly the same as that for Type II. This suggests that Types Ib and Ic come from roughly the same mass range as Type II. One possibility is that Types Ib and Ic come only from Wolf-Rayet stars that formed by the action of strong stellar winds in stars more massive than 30 solar masses (Chapter 2, Section 2). This is probably not the only source of Type Ib and Ic events. Because very massive stars are rare, there would probably be too few of them to explain the rate of explosions. This suggests that Types Ib and Ic also come from stars that were born with less than 30 solar masses. A binary companion would then be necessary to help strip away the hydrogen envelope. Nevertheless, the basic arguments that pertain to Type II supernovae hold also for Types Ib and Ic. If Types Ib and Ic come from massive stars, to account for their rate of occurrence and their sites in spiral arms, Types Ib and Ic are also very likely to be associated with core collapse to form neutron stars.

At the lower end of the mass range suspected to contribute to Type II supernovae, the evolution may be slightly different. The outcome, core collapse, is basically the same. Computer calculations show that for stars with original mass between about 8 and 12 solar masses the core will be supported by the thermal pressure when carbon is burned. This stage of carbon burning is then regulated and gentle in the standard way. The carbon burns to produce neon and magnesium, but the oxygen that typically coexists with the carbon after helium burning does not get hot enough to burn. As the core, now composed of oxygen, neon, and magnesium, contracts, the quantum pressure comes into play before any other fuel can ignite. The stars in the mass range 8–12 solar masses will therefore form cores supported by the quantum pressure and consisting of oxygen, neon, and magnesium. The atomic nuclei of neon and magnesium are capable of absorbing an electron, thus turning one proton into a neutron, and transmuting themselves into an element of lower proton number. This process reduces the electrons that are responsible for the quantum pressure that is supporting the core. The result is that the core collapses before any of the elements in the core begin thermonuclear burning. During the collapse, the remaining nuclear fuels – oxygen, neon, and magnesium – are converted to iron. The net result is a collapsing iron core, just as for the more massive stars where the iron core forms before the collapse ensues. These two processes of iron core collapse may give identical results, or there may be some subtle difference between collapse triggered by absorbing electrons rather than by heating and disintegrating the iron. These differences could affect the explosive outcome. There is some evidence that stars in the lower-mass range with the collapsing oxygen/neon/magnesium cores may be especially efficient in producing some of the rare heavy elements like platinum.

A different way of addressing the question of which stars explode is to ask which stars do not explode because they cast off their envelopes gently and leave white-dwarf remnants. This question has been addressed by counting the number of white dwarfs in stellar clusters of various ages and then estimating what stars must have produced those white dwarfs. Such estimates are roughly consistent with the statement that all stars below about 8 solar masses make white dwarfs, and hence do not make supernovae, at least not right away.

Estimates of the rate of formation of neutron stars in the Galaxy are similar to estimates of the rate of formation of Type II supernovae. This does not prove that Type II supernovae produce neutron stars, but the notion that the two processes are directly related is a nearly universal working hypothesis. The problem with this hypothesis is that no calculations have been able to satisfactorily show that the energy liberated in forming a neutron star can routinely cause an explosion. Despite rather gross changes in the physics over the last three decades, many calculations keep stubbornly predicting no explosion, but total collapse. This does not necessarily mean that such explosions do not occur in nature. The calculations may leave out some important piece of physics. That physics might be presently unknown to us, or the process might be too complex to calculate effectively, like the effects of rotation or magnetic fields. Alternatively, we may find that not all stars that develop collapsing cores do form explosions. Some may leave black holes with no explosion at all.

3. Element Factories

Stars with an initial mass larger than 20 solar masses should form iron cores that collapse. There are so few of these stars that whether they explode or not will not change the total supernova rate appreciably. Some other way must be devised to determine whether or not they explode. The observation that suggests that some of the massive stars must explode is the simple but profound one that says that about 1 percent of the material in stars is composed of elements heavier than helium. These elements cannot be produced in the big bang. Alternatively, we know from theoretical calculations that heavy elements in reasonable proportions are produced naturally in the massive stars in the process of forming an iron core. The strong suspicion is that at least some of the most massive stars must explode in order to eject their complement of heavy elements into space to be incorporated in new stars.

Calculations show that stars with mass between 8 and about 15 solar masses contain too little in heavy elements outside the collapsing core to contribute substantially to the production of elements like carbon, oxygen, and calcium, which are abundant in stars as well as in our bodies. Thus the stars that are presumed to account for most, if not all, of the Type II supernovae are not significant contributors to synthesis of the heavy elements. Stars with mass between about 15 and 100 solar masses produce substantial amounts of heavy elements. If these stars explode and eject their heavy elements, this freshly synthesized material will mix with the hydrogen in the interstellar gas. New stars form from this enriched mixture. If all stars from 15 to 100 solar masses explode, the new stars will have about the right amount of all the abundant heavy elements.

This picture has led to the widespread belief that the most massive stars must explode and produce the heavy elements. There is probably a great deal of truth in this notion. As observations get more accurate, however, there are hints that the broad picture must be reassessed. Detailed stellar spectra of both young and old stars have allowed new accurate measurements to be made of the way that various elements have been produced throughout the history of the Galaxy. There is a suggestion that if all the massive stars from 15 to 100 solar masses explode, many of the basic heavy elements like carbon, oxygen, and iron will be produced in greater quantity than is observed in the stars in the vicinity of the Sun. A possible resolution of this dilemma is that some of the massive stars collapse completely. In this picture, some massive stars would explode, ejecting heavy elements and leaving neutron stars behind as compact remnants. Others would produce no explosion and would leave behind black holes as the only remnant of their previous stellar existence.

The pattern that seems to best satisfy all our present knowledge would have stars from about 8 to about 30 solar masses exploding and those from 30 to 100 solar masses collapsing and swallowing all their heavy elements. The most reasonable position is probably to conclude that we do not yet know enough about the nuclear and evolutionary processes in stars to conclude with any certainty which stars explode and eject the heavy elements we see.

4. Collapse and Explosion

In the collapse of an iron core, the protons capture electrons and convert to neutrons. Each reaction creates a neutrino. This is the process by which the composition is converted to neutrons, the necessary step to make a neutron star. For every neutron formed, there must also be a neutrino. The result is a *lot* of neutrinos.

When the collapse reaches the density of atomic nuclei, the strong nuclear force has a repulsive component. This provides a strong outward pressure. In addition, the quantum pressure of the neutrons plays a role. The result of the increased pressure is that the collapse halts (temporarily, at least). The basic processes as they are thought to occur in a massive star are shown in Figure 6.1.

If you drop something heavy, like a bowling ball, appreciable energy is released when it lands. The more massive the object, the greater the energy released. The farther the object falls, the greater will be the energy released. Imagine dropping the bowling ball from the top of a tall building. Imagine dropping a sports utility vehicle from the top of a tall building. Now imagine the gigantic release of energy when a star with the mass of the Sun collapses to the tiny size of a neutron star, only a few kilometers across. A huge energy is released when the neutron star forms. This energy is several hundred times more than is necessary to blow off the outer layers, those containing calcium, oxygen, carbon, and helium, and any outer envelope of never-burned hydrogen. The problem is that most of the energy produced in the collapse is lost to the neutrinos that can easily stream out of the newly born neutron star and through the infalling matter. If 99 percent of the energy is lost, 1 percent can remain. That is enough to cause an explosion. If 99.9 percent is lost, however, that is too much. The explosion will fail, and the outer matter will continue to rain in and crush the neutron star into a black hole.

The exact treatment of this problem has proven to be very difficult. The requirement is to determine whether 99 or 99.9 percent of the energy is lost to neutrinos, or whether it is some fraction in between. The energy lost to neutrinos must be determined to about one part in a thousand. Uncertainties in the complex physics involved in core collapse have been larger than this critical difference. A related problem is that the explosion process tends to be self-limiting. If more of the energy is trapped, then the rate of infall of new matter from the outer parts of the star is slowed. This, however, decreases the rate at which the collapse produces energy that can power the explosion. The result has been that for decades computer calculations have tended to give results that teeter on the edge of success, some giving explosions, many giving complete collapse to form black holes with no explosion. No completely clear, accepted, reproducible, result has emerged. The

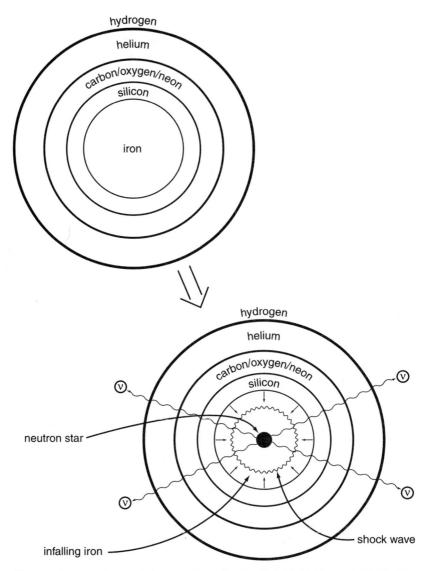

Figure 6.1. The collapse of the iron core of a massive star to form a neutron star. (top) The star passes through many phases of regulated nuclear burning and forms an iron core. (bottom) The iron core collapses to form a neutron star, momentarily leaving the outer layers hovering. The creation of the neutron star creates a huge flood of neutrinos. The rebound of the neutron star produces a shock wave that propagates outward into the infalling matter. If the shock wave is strong and the explosion is successful, the outer layers will be blown off in a supernova explosion, and the elements produced in the star will be spread into space. If the explosion is not strong enough, the outer layers will also fall in and crush the neutron star into a black hole.

stars know how to produce these explosions, but astrophysicists are struggling to figure it out.

Current research involves two basic mechanisms by which the collapse of an iron core might be partially reversed to make a supernova explosion. One is called *core bounce*. When the neutron star first forms, the new star overshoots its equilibrium configuration giving a large compression to the neutron core. There is then a rebound. This rebound sends a strong supersonic shock wave back out through the infalling matter. The core takes about 1 second to collapse after instability sets in. The core bounce creates the shock in about 0.01 second. If everything works, in this short time a huge explosion should be generated.

If the shock wave is sufficiently strong, the outer matter is ejected, and the neutron star is left behind. The shock runs uphill into the infalling matter, however. Some of the energy of the shock is dissipated by the production and loss of neutrinos. The shock also must do the work of breaking down the infalling iron into lighter elements, protons and neutrons, to form the neutron star. The shock wave can thus stall with insufficient energy to reach the outer layers of the star. Matter can continue to rain down on the stalled shock front, as illustrated in the bottom part of Figure 6.2. The shock front hangs in mid-flow, much as a bow wave stands off a rock in the middle of a stream, as shown in the top of Figure 6.2. The matter will continue to be shocked as material hits this front, but the shocked matter will continue to fall onto the neutron star, just as the water will be slowed, but not stopped, by the rock in the stream. When enough matter lands on the neutron star, the neutron star will be crushed into a black hole. Most calculations currently show that the core bounce alone is not sufficient to cause an explosion.

The other mechanism being actively considered takes advantage of the tremendous stream of neutrinos leaving the neutron star. Normal matter, the Sun, is essentially transparent to neutrinos because neutrinos interact only through the weak nuclear force. The only exception to this is neutron star matter. This matter, nearly as dense as an atomic nucleus, is so dense that it can be opaque or at least semitransparent to the neutrinos. Although most of the neutrinos will get out into space, a small fraction will be trapped in the hot matter that lies just behind the shock front created by the core bounce. The slow accumulation of neutrino heat may provide the pressure to reinvigorate the shock, driving the shock out-

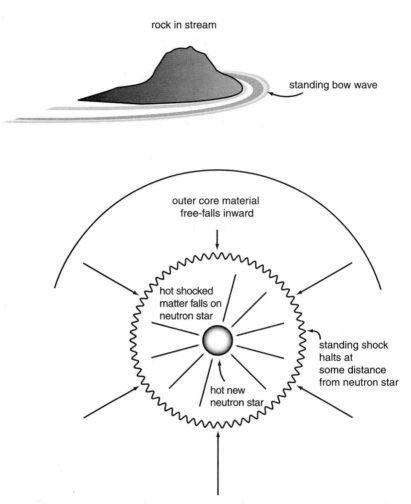

rock in stream

standing bow wave

outer core material
free-falls inward

hot shocked
matter falls on
neutron star

standing shock
halts at
some distance
from neutron star

hot new
neutron star

Figure 6.2. A rock in a stream will cause a standing bow wave to form in front of it. Because the water, not the rock, is moving, the wave can also stand still. In the collapse of a stellar core, the shock wave formed by the rebound of the neutron star will move outward into the infalling matter. It can reach a position where the pressure of the hot gas inside the shock (the analog of the rock) supports the shock as the outer matter of the star continues to rain downward. As the matter flows inward, the shock can hover at one radius as a "standing shock."

ward and causing the explosion. Slow in this case means about a second.

The mechanism for depositing a small fraction of the neutrino energy behind the shock may be related to the boiling of the newly formed neutron star, as shown in Figure 6.3. When the collapse is

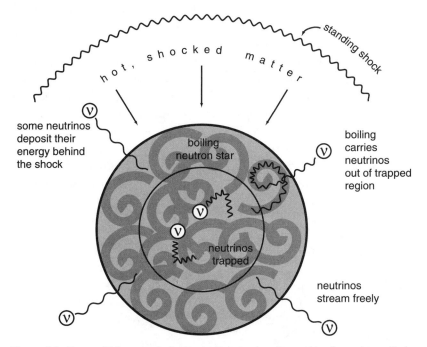

Figure 6.3. Deep within a newly formed neutron star, the matter is so dense that even the neutrinos are trapped. The neutrinos can bounce around, but they cannot escape directly. If the neutron star is hot and boiling as computer calculations show, some of the matter containing neutrinos will boil to the surface, where the trapped neutrinos can escape. This can enhance the rate of loss of neutrinos from the neutron star. Some fraction of these neutrinos can interact with matter beyond the neutron star, but behind the standing shock. If the flood of neutrinos enhanced by the boiling of the neutron star is large enough, sufficient neutrino energy might be deposited behind the standing shock to reinvigorate it and send it all the way out of the star, leading to a successful supernova explosion.

first halted and the neutron star rebounds, the neutron star is very hot. This heat can cause the neutron star to boil much like a pan of water boils on the stove. The boiling provides a mechanism for carrying the heat upward, in the case of the pan, or outward, in the case of the neutron star, by mechanical motion that bodily carries the heat. Under the right circumstances, this boiling process can be much more efficient in transporting heat than a slower process of leaking radiation, or neutrinos. Calculations of this process in neutron stars are very challenging because the motion is complex. All modern calculations that can follow motion in more than one (radial) dimension show that neutron stars do boil. There is a consensus that explosions will not occur without this boiling. There is still debate about whether this process is sufficient to cause an explosion.

Most of the current calculations of core bounce and subsequent events treat the configuration as spherically symmetric. Even if the neutron star boils, the structure of the neutron star may, on average, be spherically symmetric. There are a number of lines of evidence, however, that the explosions that result from the core collapse process are intrinsically nonspherical. Matter may be ejected more intensely in some directions than others. If this is the case, then the current numerical calculations may be missing a major ingredient necessary to yield an explosion. The most obvious mechanism for breaking the spherical symmetry by singling out a specific direction is rotation because rotation defines a rotation axis. Proper treatment of rotation, perhaps abetted by magnetic fields, may be necessary to understand fully when and how collapse leads to explosions. Adding the effects of rotation and magnetic fields is even more of a computational challenge, but computer power grows steadily, and progress will be made in this area in the next few years. New suggestions that rotation and magnetic fields are important to this process are presented in Chapter 11.

5. Type Ia Supernovae: The Peculiar Breed

The principal peculiarity of Type I supernovae is that such events have no hydrogen in their ejected material. The hydrogen envelope that surrounds most stars has either been ejected or consumed to helium or heavier elements. As noted in Section 1, there are two rather different observed categories of Type I. Some of them, the Types Ib and Ic, like Type II, occur only in spiral or irregular galaxies. The Type Ia supernovae occur in all types of galaxies. This makes Type Ia events different in some fundamental way and worthy of special attention.

In particular, Type Ia supernovae occur in elliptical galaxies, whereas Types II, Ib, and Ic do not. Elliptical galaxies have converted essentially all their gas into stars long ago and to a great extent have ceased the making of stars. Thus elliptical galaxies are thought to consist of only old, low-mass, long-lived stars. The high-mass stars born long ago should be long dead. This has given rise, in turn, to the idea that Type Ia supernovae must come somehow from low-mass stars. Because spiral galaxies contain a mix of high-mass and low-mass stars, that spirals produce both Type Ia and Type II supernovae is not surprising.

Another aspect that has driven thinking about Type Ia supernovae is that their observed properties are remarkably uniform. Type Ia events tend to follow the same light curve. In addition, as Type Ia brighten and decline, the alterations in their spectra follow a very predictable course. Because white dwarfs of the Chandrasekhar mass would be essentially identical and hence undergo nearly identical explosions, the observed homogeneity of Type Ia has pointed to an origin in exploding white dwarfs. We now know that all Type Ia supernovae are not exactly identical. The reasons for this are the subject of active current research, as will be discussed later.

The most popular notion for how to turn a low-mass star into a supernova is thus to rejuvenate a white dwarf. The idea is that the more massive star in an orbiting pair could evolve and form a white dwarf. The low-mass companion could then take a long time to evolve, but the companion would eventually swell up as a red giant and dump mass onto the white dwarf. If the total mass accumulated by the white dwarf approaches the Chandrasekhar mass of about 1.4 solar masses, the white dwarf might then explode. A variation on this theme is that the white dwarf could grow in mass in a cataclysmic variable system where the mass flows from a main sequence star (Chapter 5). This process is slow, and the system could still last a long time before exploding. Yet another possibility is that Type Ia supernovae arise from systems of two white dwarfs that slowly merge due to the emission of gravitational waves generated by their orbital dance.

Careful studies of the observed properties of Type Ia supernovae are completely consistent with the general picture that the explosion occurs in a white dwarf. Near peak light, the spectra of Type Ia supernovae show elements such as oxygen, magnesium, silicon, sulfur, and calcium. These are just the elements expected if a mixture of carbon and oxygen burns to produce somewhat heavier elements consisting of differing numbers of "helium nuclei." As a Type Ia supernova evolves, the spectrum becomes dominated by iron and other similarly heavy elements. These elements can be produced by burning carbon and oxygen all the way to iron. The nuclear binding energy of iron is at the bottom of the "nuclear valley," where the neutrons and protons in the nucleus are most compressed.

In the process of expanding and thinning out, the outer, more tenuous portions of a supernova are seen first, and the inner, denser, more opaque portions are only seen later. The information revealed by the evolution of the spectra is then consistent with a configuration in which the denser inner portions of the exploding star burn all the way to iron and ironlike elements, but the outer parts are composed of matter that results from carbon burning, but that is not so thoroughly processed. Computer models of exploding white dwarfs give results that match this pattern rather well. The exact nature of the combustion is still being explored, but the most successful models adopt a progenitor that is a carbon/oxygen white dwarf with a mass very near to, but less than, the Chandrasekhar mass.

At this point, I must correct a long-standing and erroneous view of the nature of Type Ia supernovae. This view is shared by many wise experts and neophytes alike because they have not followed this research closely. A casual view that permeated the astronomical community and the popular astronomical literature decades ago, and that is very difficult to root out, is that to make a Type Ia supernova, matter is added to a white dwarf until the Chandrasekhar mass is exceeded and the white dwarf collapses. *This is wrong!* The reason this notion is so persistent, I suspect, is that the idea of exceeding the mass limit and collapsing is simple and visceral. In addition, the "other" means of making supernovae does involve core collapse, and so it is easy to confuse the two mechanisms. There are also circumstances where some white dwarfs might collapse, but if so, the process does not yield the events we observe as Type Ia supernovae. Rather, mass is added, we believe, increasing the density in the center of the white dwarf until finally carbon can ignite. This condition of carbon ignition and subsequent unregulated thermonuclear runaway happens when the white dwarf has a mass about 1 percent less, not more, than the Chandrasekhar mass, and it blows the white dwarf up completely, so there is no collapse. This is a somewhat more complicated and perhaps less intuitive process (think dynamite!), and this may be why it has not permeated all corners of the community of interested people. Nevertheless, the supernova community stopped talking about exceeding the Chandrasekhar limit and collapse in the 1960s, and it is rather dismaying to find experts in related areas, never mind popular astronomy enthusiasts, still referring to this outmoded physical picture. The overwhelming observational evidence is that Type Ia supernovae arise from carbon/oxygen white dwarfs of mass a little less than the Chandrasekhar limit that do not collapse, but blow up completely by a process of thermonuclear explosion.

Type Ia supernovae explode because the white dwarf is supported by the quantum pressure, and any burning under those circumstances is unregulated, as we discussed in Chapters 2 and 5. For Type Ia supernovae, burning is unregulated in the extreme. As a white dwarf approaches the Chandrasekhar limiting mass, the central density gets very high. Formally, the density would go to infinity just at the Chandrasekhar limit, but in practice other physics, in this case carbon burning, will come into play. The high density triggers the ignition of carbon but also ensures that, under these circumstances, the quantum pressure will be exceedingly large. The white dwarf will have a finite temperature that will help to promote the carbon burning, but the thermal pressure is negligible. The story of unregulated burning we have told before will then play out in the most drastic way. The carbon begins to burn and to release energy. The quantum pressure does not budge. There is no mechanical response to expand and cool the star and damp the burning. The burning goes even faster, raising the temperature even more and producing ever faster burning. Under the extreme conditions at the center of a white dwarf with a little less than the Chandrasekhar mass, the burning cannot be controlled, the oxygen also ignites, and all the fuel is consumed to iron-peak elements in a flash. The result is a violent thermonuclear explosion.

There are two different ways of propagating a thermonuclear explosion in a white dwarf. One is a subsonic burning like a flame, a process called a *deflagration*. The other is a supersonic burning that is preceded by a shock front, much like a stick of dynamite. This process is known as a *detonation*. The most sophisticated current models, those that best match the data, have the unregulated carbon burning begin as a boiling, turbulent deflagration and then make a transition to a supersonic detonation, as illustrated in Figure 6.4. Such models naturally create ironlike matter in the center and intermediate elements like magnesium, silicon, sulfur, and calcium on the outside. These models also predict that the white dwarf is completely destroyed, leaving no compact remnant like a neutron star or a black hole. This comparison of theory and observation thus strongly points to an interpretation of Type Ia supernovae as the explosion of a carbon/oxygen white dwarf at just less than the Chandrasekhar limit.

This does not, however, answer all the mysteries about the nature of Type Ia supernovae. For Type II supernovae, we think we under-

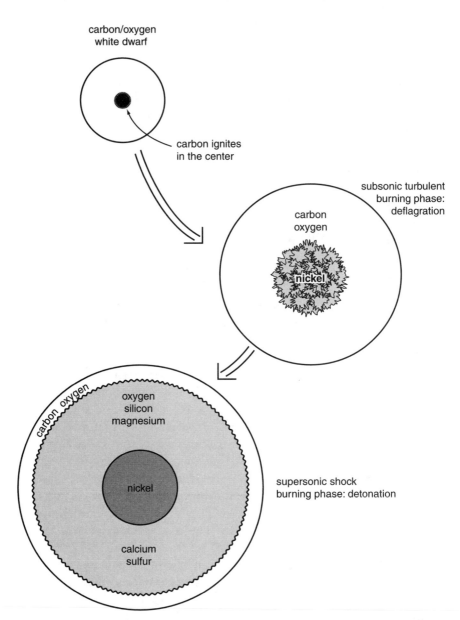

Figure 6.4. (Top) A Type Ia supernova explosion begins with the ignition of carbon near the center of the white dwarf. (Middle) A turbulent, roiling burning front that moves less rapidly than the speed of sound spreads out from the center, at first converting all the burning matter to radioactive nickel. The pressure waves from this burning cause matter beyond the burning regions to expand before the burning reaches them. (Bottom) At some point, the burning front begins to propagate super-sonically, producing a shock wave that triggers the burning. This detonation wave moves so rapidly that the outer portions of the star cannot expand substantially farther before they are overtaken by the burning. The detonation burning leaves behind oxygen, magnesium, silicon, sulfur, and calcium, the elements seen in the outer layers of Type Ia supernovae. A thin layer of unburned carbon and oxygen on the outside of the white dwarf might survive the explosion.

stand the broad outlines of the evolution of massive stars to form collapsing iron cores. We do not understand how the collapsing core results in an explosion. For Type Ia supernovae, the situation is just the opposite. There is nearly unanimous agreement that the mechanism of Type Ia supernovae is a violent thermonuclear explosion that obliterates the star. Despite this convergence of opinion on the mechanism, there is no generally accepted picture of the evolutionary origin of these peculiar events. The question of how the white dwarfs grow to the Chandrasekhar mass is still a knotty, unsolved problem. There is no direct evidence that Type Ia supernovae arise in binary systems. Despite this lack of direct evidence, all the circumstantial evidence points to evolution in double-star systems, and there are few credible ways of making a white dwarf explode without invoking a binary companion. The challenge is to figure out what binary evolution leads to a Type Ia explosion.

The task of figuring out the prior evolution of Type Ia supernovae is made harder if one accepts that the supernovae arise in white dwarfs of the Chandrasekhar mass. Recall from Chapter 5 that the average white dwarf has a mass of only 0.6 solar mass. This means that the mass must more than double if the process starts with one of these white dwarfs. The task might be made easier if the white dwarfs born in binary systems are systematically more massive. There is some evidence that this may be the case. Note that if the white dwarf is in a system that undergoes a nova explosion every 10,000 years or so, the mass of the white dwarf could actually decrease! This is not an easy problem.

For this reason, there has been considerable attention paid to mechanisms that would lead a white dwarf to explode even though it had less than a Chandrasekhar mass. The most likely such model is one where a white dwarf accretes mass rapidly enough that the accreted hydrogen remains hot and supported by its own thermal pressure. The hydrogen then burns on the surface of the white dwarf in a regulated manner, and a nova explosion is avoided. Under these circumstances, however, a thick layer of helium can build up surrounding the inner carbon/oxygen core. The helium layer can be supported by the quantum pressure. If this helium ignites, computer models show that a violent explosion occurs. The explosion not only burns the helium but can send a shock wave inward that causes the inner carbon/oxygen white-dwarf core to burn as well. All this happens very quickly, a matter of seconds, so the result is a single powerful explosion. This is a very plausible mechanism to produce an explosion. The problem is that this mechanism does not produce results that are in good agreement with the observations. The helium burns to ironlike material on the outside that should be seen first and produces only thin layers of intermediate elements like silicon and calcium that are ejected with the wrong velocities. The ejecta tend to be too hot as well. Despite the appeal of these models, nature seems to prefer exploding white dwarfs of nearly the Chandrasekhar mass.

There are currently two "best bets" for how to generate Type Ia supernovae. Both involve mass transfer onto a white dwarf in a binary system. One invokes transfer of hydrogen from a red giant at just the right rate. The mass transfer must be rapid enough that the collected hydrogen does not undergo a nova explosion that ejects the hydrogen along with part of the white dwarf. Apparently, the mass transfer must be rapid enough that even the helium remains hot, supported by the thermal not the quantum pressure, so that igniting the helium does not cause an explosion with the wrong properties. If the mass transfer is too rapid, however, a common envelope of hydrogen will engulf the white dwarf. The hydrogen should show up in the explosion. That would be a violation of the basic observational definition of a Type I supernova. There may be binary configurations where the mass transfer is "just right." The hydrogen will burn gently to helium, the helium will burn gently to carbon and oxygen, and that carbon and oxygen will settle onto the core to cause the core to grow toward the Chandrasekhar mass. Candidate systems have even been identified among a special class of X-ray sources called *supersoft X-ray sources*. Unfortunately, careful modeling of these systems has failed to provide convincing evidence that there are enough of them to account for the rate at which Type Ia supernovae are observed to explode. In addition, recent work has suggested that the process of burning the helium can never be completely gentle and that no process of transferring hydrogen will work to produce a Type Ia.

The other popular model for producing a Type Ia supernova is by the merging of two white dwarfs in a binary system. This merging must happen sometimes. Some binary white dwarfs are seen. There is still controversy concerning whether there are enough binary white-dwarf systems with total mass exceeding the Chandrasekhar mass to produce Type Ia supernovae at the observed rate. In addition, the process by which the smaller-mass white dwarf fills its Roche lobe and comes apart, dumping its mass on the larger-mass white dwarf as described in Chapter 5, is complex and not well understood. The disrupted matter will swirl around the larger-mass white dwarf in a thick disk. How that matter will settle onto the remaining white dwarf is not completely clear.

New perspectives on the nature of Type Ia supernovae came with evidence produced in the 1990s that confirmed a long-standing suspicion. Type Ia supernovae are not all identical. They show interesting variations that are mostly subtle, but real. In some cases, the variations are not even so subtle. The general trend is that Type Ia supernovae that are brighter than average decline from maximum brightness a bit slower than average. The events that are a bit dimmer than average (some by as much as a factor 2) decline more rapidly. Models of exploding Chandrasekhar-mass white dwarfs can account for this behavior if the explosion in some stars makes the transition from a subsonic deflagration to a supersonic detonation a little earlier than in others. Why this should be so is the object of current research.

The observed variety of Type Ia behavior seems to correlate with the nature of the host galaxy. Elliptical galaxies seem to produce selectively Type Ia supernovae that are of the dimmer, more rapidly declining variety. Within spiral galaxies, the inner portions seem to produce the full range of behavior, but the outer parts of the galaxy produce especially homogenous explosions. We do not yet understand all the variables, but there is probably a variety of ways of making white dwarfs explode, and the progenitor systems can display a range in ages. Some Type Ia supernovae may come from mass transfer in "normal" binary systems, from some variation on a cataclysmic variable. Others may come from merging white dwarfs. Some may come from stars near 8 solar masses that have relatively short lifetimes and others may come from stars with closer to 1 solar mass that have lifetimes approaching that of the Universe itself.

6. Light Curves: Radioactive Nickel

Supernovae display a variety of shapes to their light curves. Type Ia supernovae are the brightest. They decay fairly rapidly in the first 2 weeks after peak light and then more slowly for months. Some Type II supernovae have an extended plateau and some drop rather quickly from maximum light. Both types seem to have a very slow decay at very late times, several months after the explosion. Types Ib and Ic supernovae are typically fainter than Type Ia by about a factor of 2, but they have similar shapes near peak light and show evidence for a slow decay at later times. These patterns tell us something about the star that exploded and about a fundamental process that is probably taking place in all of them: radioactive decay.

When a supernova first explodes, the matter is compact, dense, and opaque. To reach maximum brightness, the ejected matter must expand until the material becomes more tenuous and semitransparent. The size the ejecta must reach is typically 10,000 times the size of the Sun. This is 100 times the size of a red giant and 100 times the size of the Earth's orbit. As the matter expands, however, it cools. If the matter must expand too far before heat leaks out as radiation, the material may have cooled off so that there is no more heat to radiate.

Most Type II supernova explosions are thought to occur in red-giant envelopes. These are very large structures. After the explosion, large envelopes do not have very far to expand before they become sufficiently transparent to leak their heat as light. As they begin to radiate, Type II supernovae still retain a large proportion of the heat that was deposited by the shock wave that accompanied the super-nova. Near maximum light and on the typical plateau that lasts for months, Type II supernovae shine by the shock energy originally deposited in the star. The deposited energy presumably arises in the core collapse process.

For a Type I supernova, however, the story is different. Whether the exploding star is a white dwarf, as suspected for a Type Ia, or the bare core of a more massive star, as suspected for Types Ib and Ic, the exploding object is very small. The expected sizes range from one-tenth to one-thousandth of the size of the Sun. These bare cores are vastly smaller than the size to which they must expand before they can leak their shock energy. The result is that the expansion strongly cools the ejected matter, and by the time the matter reaches the point where it could radiate the heat, the heat from the original shock is all gone. This kind of supernova requires another source of heat to shine at all. All the light from Type I supernovae comes from radioactive decay.

The nature of a thermonuclear explosion is to burn very rapidly. If the explosion starts with a fuel built from multiples of helium nuclei – carbon, oxygen, or silicon – that has equal numbers of protons and neutrons, then the immediate product of the burning will also have equal numbers of protons and neutrons. This is because the rapid burning takes place on the time scale of the strong nuclear reactions. To change the ratio of protons to neutrons requires the weak force and thus a longer time. Nature, however, does not leave the burned matter with equal numbers of protons and neutrons. Rather, nature prefers to form the element with the most tightly compacted nucleus, that of iron, which has twenty-six protons and thirty neutrons.

Nature manages to make iron in a thermonuclear explosion in a three-step process. The first step is to forge an element that is close to iron but that has equal numbers of protons and neutrons. This element, like iron, has a nucleus that is tightly bound by the nuclear force and has the same total number of protons plus neutrons, fifty-six, but with twenty-eight protons and twenty-eight neutrons. This is the element that will form first, before the slower weak interactions come into play. This condition singles out one element, nickel-56. The unregulated burning of carbon or oxygen or silicon will naturally first produce nickel-56.

Nickel-56 is, however, unstable and therefore undergoes radioactive decay. The radioactive decay is induced by the weak force. One of the protons in the nickel converts to a neutron. The result is the formation of the element cobalt-56 with $28 - 1 = 27$ protons and $28 + 1 = 29$ neutrons. In the process, an electron is absorbed to conserve charge, and a neutrino is given off to balance the number of leptons. Excess energy comes off as gamma rays, high-energy photons. The gamma rays can be stopped by collision with the matter being ejected from the supernova and their energy used to heat the matter. The hot matter shines as the light we observe on Earth. The power of the light falls off as the nickel decays away and as the matter expands so that it is less efficient in trapping the gamma rays. The neutrino always just leaves the star and plays no role in this heating.

The cobalt-56 that forms is also unstable. Again, the weak force induces a proton to convert to a neutron. The result has $27 - 1 = 26$ protons and $29 + 1 = 30$ neutrons. This is just good old iron-56, nature's ultimate end point. This decay again produces a neutrino and gamma-ray energy. In this case, charge is conserved by emitting an antielectron, or positron. The positron will quickly collide with one of the electrons that are floating around normally, one for every proton. The annihilation of the electron will produce another source of gamma rays. Iron-56, with twenty-six protons and thirty neutrons, sits at the bottom of the nuclear energy valley, and so it is stable. This radioactive decay scheme, nickel to cobalt to iron, is just one of nature's ways of rolling things down the nuclear hillside to become iron.

The radioactive decay of these elements is controlled by a quantum uncertainty. One does not know what atom will decay, but on the average half will decay in a given time. For nickel-56, the time for half to decay is 6.1 days. After another interval of 6.1 days, half of the remaining half will decay so that after 12.2 days only one-quarter of the original nickel remains. After 18.3 days, only one-eighth of the original nickel will survive. This time scale, about a week, is the time for the gamma rays from the radioactive decay to pump energy into the exploding matter. Likewise, the cobalt-56 decays with a half-life of about 77 days, roughly 2 months. These times are long compared with the times for the basic explosion to ensue, a matter of seconds. That is why the nickel-56 forms first in this type of explosion and the iron forms only later, over several months. The observed light curves of Type I supernovae decay somewhat faster than the decay of nickel-56 in the early phase and of cobalt-56 in the later phases. The reason is that not all the gamma rays produced in the decay are trapped and converted to heat and light. Some of the gamma rays escape directly into space.

For Types Ib and Ic, the amount of nickel required to power the light curve is about one-tenth of the mass of the Sun. This amount of nickel is consistent with many computations of iron-core collapse. The nickel is produced when the shock wave, of whatever origin, impacts the layer of silicon surrounding the iron core. Type Ia supernovae are generally brighter and must produce more nickel, of order 0.5–1 solar mass. The dimmest Type Ia events require only 0.1–0.2 solar mass of nickel. The models of Type Ia supernovae based on thermonuclear explosions in carbon/oxygen white dwarfs of the Chandrasekhar mass produce this amount of nickel rather naturally in the explosion. The amount can vary depending on, for instance, the density at which the explosion makes the transition from a deflagration to a detonation, so the variety of ejected nickel mass can also be understood, at least at a rudimentary level.

If Types Ib and Ic are related to the cores of massive stars, as the circumstantial evidence dictates, then their explosion mechanism should be similar to that of Type II supernovae. This suggests that Type II should also eject about 0.1 solar mass of nickel-56. This is not enough to compete with the heat and light from the shock near maximum light, but as the ejected matter continues to expand and cool, the shock energy dissipates, and the supernova gets dimmer. At this phase, the dimmer but steady source of radioactive decay should take over. The evidence from fading Type II supernovae shows that this is the case. Once again, not all the gamma rays are trapped. Some must radiate directly into space. A properly designed gamma-ray detector flown in orbit should see these missing gamma rays and directly confirm the validity of this picture. As we will see in Chapter 7, this was the case for SN 1987A.

WHEN BETELGEUSE BLOWS

For years, every time I gave a popular lecture on supernovae, someone would ask, "What will happen to the Earth when a nearby supernovae explodes." Each time I would say, "I thought about that a little a long time ago, but I really need to work that out, so I know how to answer this question." Then after the lecture, I would return to work-a-day issues and forget until the next popular lecture. To get a record down on paper that I can use in the next lecture, here is a sketch of what will happen when the most likely nearby star explodes.

Betelgeuse is a red-giant star that marks the upper-leftmost shoulder of the constellation of Orion as we look at it from Earth. You can see it easily from anywhere in the northern hemisphere on a winter or spring evening. We do not know the precise mass of Betelgeuse, but we can make an intelligent guess. That will give us a good guess as to its fate and what will happen at the Earth.

Thanks to careful measurement by triangulation we know quite accurately how far away Betelgeuse is. It is 427 light years away. That is long by human standards, but right next door in a Galaxy that is 100,000 light years across. There are closer stars, but none that are likely to explode. At this distance, Betelguese presents little threat to the Earth, but we will sure notice it when it goes off. It is a good example of the low-level impact that will contribute to the stochastic history of bombardment of the Solar System by astronomical events over its 5-billion-year history. Such events should occur roughly once every million years.

From the power received at Earth over all wavelength bands and its distance, we can estimate that Betelgeuse emits a luminosity of about 50,000 to 100,000 times that of the Sun. From computer models, we can further estimate that this luminosity in a red giant requires a star of original main sequence mass of about 15–20 solar masses. This mass is such that, in the absence of a stellar companion, and Betelgeuse seems to have none, there will be little mass loss to winds, so this is probably a pretty good estimate. Stars in this mass range are predicted to evolve iron cores and undergo

Betelgeuse produces a bright pulsar (Chapter 8), it might be a substantial source of gamma rays for thousands of years.

The ejecta from Betelgeuse will freely expand for about 1,000 years and span about 20 light years in that time. During this time, the ejecta will be cold and dim. The supernova material will then start to pile up appreciable mass in interstellar matter and enter the supernova remnant phase. The supernova remnant will turn on as an X-ray source and begin to produce cosmic rays by acceleration of particles at the shock front. The supernova material will slow down, but a shock will race ahead into the interstellar matter, decelerating as it sweeps up ever more mass. The shock wave in the interstellar matter will be fully developed in about 20,000 years when it has expanded to about 30 light years. The shocked matter will begin to radiate substantially and cool off when it has expanded to about 100 light years about 100,000 years after the explosion. The remnant will plow on through the interstellar matter. The shock from Betelgeuse will be very mild by the time it reaches the Solar System and will probably be easily deflected by the solar wind and magnetopause. The exception might be if there is a low-density, interstellar "tunnel" between us and Betelgeuse that would channel some of the energetic matter to us before it slowed down.

All these effects would be much stronger if the supernova were only 30 light years from the Earth. There are no candidate stars around us now, but on its galactic journey, such nearby explosions have probably happened several times in the 5 billion year life of the Earth. Such events could be dangerous by triggering harmful mutations, but they might also be helpful because evolutionary "shocks" can also single out healthy mutations and drive biocomplexity. The Earth is coupled to this complex galactic environment, and the story of life on Earth will not be fully known until such long-term, sporadic effects are understood.

7

Supernova 1987A

Lessons and Enigmas

1. The Large Magellanic Cloud Awakes

The first supernova discovered in 1987 turned out to be the most spectacular supernova since the invention of the telescope. SN 1987A was the first supernova easily observable with the naked eye since the one recorded by Kepler in 1604. This event also brought the first direct confirmation that our basic picture of the exotic processes that mark the death of a massive star is correct. SN 1987A is the best-studied supernova ever, but the story is still unfolding, and there is much to learn.

SN 1987A did not explode in our Galaxy, but in a nearby satellite galaxy to our own Milky Way galaxy. This satellite galaxy cannot be seen from the northern hemisphere. The first European to record it was Magellan during his epic attempt to sail around the world. In English, it carries the name the Large Magellanic Cloud for this reason. People native to the southern hemisphere were undoubtedly familiar with it before that. The Aborigines living around Sydney had long had another name for it: Calgalleon, which had to do with a woolly sheep. The Large Magellanic Cloud has a somewhat smaller companion that has picked up the unimaginative name Small Magellanic Cloud. In the same Aboriginal dialect, it was rendered Gnarrangalleon. There is poetry!

The Large Magellanic Cloud is only 150,000 light years away, as shown in Figure 7.1. This is not much farther than the span across the Milky Way itself, about 50,000 light years. By contrast, the Andromeda galaxy, Messier 31, the great sister spiral galaxy to the Milky Way in our local group of galaxies, is about 2 million light years away. The nearest rich cluster of galaxies that has provided many well-studied supernovae in the last several decades is about 50 million light years away. The most distant supernovae ever found are more than a billion light years away. The nearness of the Magellanic Cloud was responsible for the great apparent brightness of SN 1987A. Intrinsically, it was relatively dim as supernovae go.

The known distance to the Large Magellanic Cloud gives us another perspective. The supernova actually exploded about 150,000 years ago, before modern *Homo sapiens* walked the Earth. By an incredible piece of luck, the light arrived at Earth just as our science had developed to the point where we could read many of its most

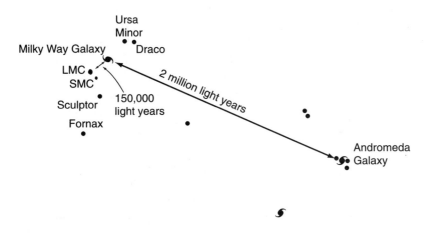

Figure 7.1. A schematic sketch of some of the twenty-one galaxies known to exist in the local group. These galaxies are distributed in three dimensions. This perspective corresponds to looking approximately along the plane of our Galaxy. The great Andromeda spiral galaxy is about 2 million light years away. By contrast, the Large and Small Magellanic Clouds are very nearby. The 150,000 light years to SN 1987A in the Large Magellanic Cloud was not much farther than one end of our Galaxy is from the other.

important messages. We had to crawl out of our caves, invent fire and the wheel, develop agriculture and writing, and witness the flowering of Greece, the Middle Ages, the Renaissance, and the Industrial Revolution. We had to develop modern science, quantum theory, Einstein's theory, an understanding of the way stars work, and the techniques for detecting neutrinos and get all this done before the light arrived! Whew!

On the other hand, if the supernova had been a mere 100 light years farther away, technology would have advanced, and we might have learned vastly more from it. On a personal note, if I had known that the light from the supernova were encroaching on the orbit of Pluto in September of 1986, I might not have agreed to be the Chair of my department that fall. By the next spring, I felt as if I were trying to drink from two fire hoses at once.

The Large Magellanic Cloud is neither a spiral nor an elliptical galaxy. Rather it is classified as an irregular galaxy. It has a large central band of rather young, newly formed stars, but then a more distended array of older stars. Off to one side of the central band, there is a region of especially intense recent star formation. The highlight of this region is called 30 Doradus by astronomers, or the Tarantula nebula by star gazers for the "hairy" arms of gas that extend from the center. The 30 Doradus region contains a very young cluster of very massive stars, perhaps 100 solar masses apiece. Surrounding the middle of 30 Doradus are large patches of gas and dust and other young massive stars, somewhat older than the core cluster of 30 Doradus. By careful study of the stellar ages, astronomers have been able to track propagating swaths of star formation in the region. One of the stars left behind in a prior wave of star formation became SN 1987A. Despite the obvious evidence for ongoing star formation, the Large Magellanic Cloud is relatively immature in the sense that it has not processed as much of its gas through stars as has the Milky Way. The amount of heavy elements in the Large Magellanic Cloud is only about one-quarter of that in our Sun.

2. The Onset

SN 1987A was discovered and first formally reported on February 23, 1987, by Ian Shelton, a graduate student from the University of Toronto who was using a small telescope at the Las Campanas Observatory high in the Chilean Andes. The first person to notice it may have been one of the night assistants, Oscar Duhalde, a Chilean of Basque extraction (Figure 7.2). Oscar had worked on the mountain for years and was justifiably proud of his familiarity with the southern

Figure 7.2. Photo of the author and Oscar Duhalde at the site of Ian Shelton's original discovery at Las Campanas Observatory at the time of the tenth anniversary of the discovery of SN 1987A. (Photo courtesy of the author.)

sky. He stepped out of the dome for a cigarette and looked at the Large Magellanic Cloud. He noticed that there was a new light in 30 Doradus but did not remark to anyone at the time about it. The supernova was still faint at the time, only hours old, and Duhalde's note of it remains one of the remarkable parts of the story. Half a world away in Australia, Rob McNaught was working on his routine survey of the sky for asteroids. He was especially tired that evening and went to bed without developing his plates. He awoke the next day with the astronomical world full of news of Shelton's announce-

Figure 7.3. Series of photos of SN 1987A taken by Rob McNaught. The first was taken on February 22, 1987, the day before the supernova. This photo shows the broad central band of newly formed stars in the Large Magellanic Cloud. The entire galaxy is much bigger than the scale encompassed by this photo. In the upper middle is the Tarantula nebula or 30 Doradus, and to the lower right of that the supernova is in the final stages of silicon burning and near to undergoing core collapse. The second photo was taken on February 23, when the supernova was only hours old. The neutrinos were long gone, but the shock wave had only recently broken through the outer layers of the star, and the supernova was brightening rapidly. This was when Oscar Duhalde noticed it. The next photo is from February 24, when the supernova was a day old. By this time, Ian Shelton had made his discovery, and the world was awakening to the amazing event. The next photo was taken on May 20 when the supernova was near maximum brightness. The image of SN 1987A does not look much brighter than the other photos because the exposure was shorter. Note that the main bar of stars and 30 Doradus look fainter in contrast and the supernova stands out clearly. The last photo was taken on August 23, as the supernova was fading. This is the time when I saw the supernova (see sidebar) and received this precious set of slides from Rob McNaught. (Photos by Rob McNaught.)

ment and found, when he did develop his image, that he had the first permanent recording of the light from the supernova. Who knows how many other people might have seen something and not mentioned it. There were rumors, but none were confirmed. Figure 7.3 shows a series of photos taken by McNaught with his patrol camera as SN 1987A appeared, brightened, and dimmed over the course of several months.

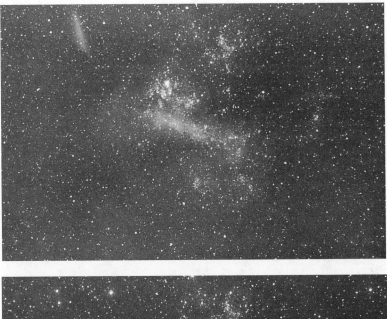

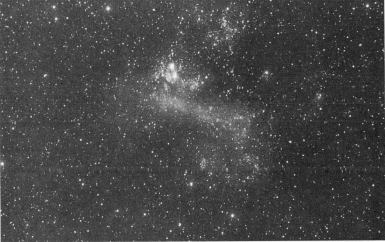

SEEING SN 1987A

I was one of the first people to hear about the supernova in the northern hemisphere. One of our ex-graduate students, Marshall McCall, was at the University of Toronto when the news came in from Ian Shelton. Marshall promptly called me. I called Nino Panagia who had used the *International Ultraviolet Explorer* satellite to study previous supernovae. Then I called Bob Kirshner at Harvard, perhaps the preeminent supernova observer of the time. I think Bob has never quite forgiven me for calling him second. Bob was also suspicious because I had been around at a meeting in Sicily in 1978 when a wonderful prank was played on him, pretending to bring news of a supernova in Andromeda. I was completely uninvolved in that prank, but guilty by association. Bob's first reaction was that I was pulling his leg. After my call, he went down the hall to the Center for Astronomical Telegrams and found their teletype spewing news of the supernova, although no one had bothered to tell him. I think he was irritated at that, too.

One of my first reactions to the supernova was to try to think of a way to go see it. This was reinforced by one of my colleagues, Don Winget, who said, "Craig, you will die a bitter old man if you don't see this supernova for yourself." Upon more reflection, I decided that I could be of more use by staying in Austin and trying to contact as many people as possible in the southern hemisphere to alert them to the event and helping to guide observations. I am not an observer myself. I did have some experience in trying to coordinate observations of supernovae at McDonald Observatory and few observatories at the time had any experience in observing supernovae.

One of the first things I did was to consult with Brian Warner, an astronomer visiting Austin from South Africa. We communicated with his colleagues who were beginning to make observations. One of the things I had learned was that if one looked at crude data when it first comes off the telescope, there was some danger of mistaking the strong spectral line of hydrogen that is prominent in Type II supernovae with the strong silicon line that is characteristic of Type Ia supernovae. Some people had mistaken Type Ia for Type II on this basis. I tried to issue this caution to my South African colleagues. They had data showing excess emission in this tricky region of the spectrum. I merely meant to be careful in the identification when they said they thought it was hydrogen. Somehow this came across in the tense rush of those first few hours as a statement that their feature was not hydrogen, but silicon, and that they were looking at a Type Ia. They announced that. Meanwhile other astronomers had done a quick and dirty analysis and recognized that they were, indeed, looking at hydrogen and announced, correctly, that SN 1987A was a variety of Type II supernova. I think some of the South Africans still hold a mild grudge against me for that.

I also thought that the supernova might emit X-rays. A few supernovae had done so, but there was no clear understanding of the mechanisms and timing of the X-rays. It did seem that if there were going to be X-rays, it was important to look very early in the explosion when the ejected matter was hot and bright. I called Walter Lewin, an X-ray astronomer at MIT. Walter pointed out that the Japanese had just launched a new X-ray satellite called *Ginga*, meaning galaxy in Japanese. Walter said that I should call Prof. Minoru Oda, the scientist who was the head of the *Ginga* team. I looked at my watch and we did a quick calculation. It was one in the morning in Tokyo. Walter said, "If I were you, I would call him." I noticed that Walter did not volunteer himself to make the call. I decided, what the heck, once in 400 years, it was worth the disruption. I got Oda's home number from Walter and rang him up. His wife answered, very sleepy, but very polite. I have the feeling she had handled emergencies before, if not one quite like this. She put Prof. Oda on the phone, and I tried to explain the circumstances as best I could. No one could be sure the supernova was producing X-rays, but looking at it with *Ginga* was the only way to find out. Prof. Oda thanked me and hung up. I heard years later that Prof. Oda had his own version of this story of "some crazy American calling him in the middle of the night." Fortunately, he did not remember who it was. As it turned out, there were no X-rays to be seen in those first few days, so I could have waited until it was a civilized time in Tokyo to call. *Ginga* did see X-rays a few months later, a detection that revolutionized some of our ideas about the supernova.

I did get a chance to see the supernova myself. Our Japanese colleagues added the topic of SN 1987A to a previously scheduled meeting in Tokyo in August of 1987, which was 6 months after the discovery. The reasonable thing to do seemed to be to go to Tokyo by way of Australia. I went with my colleague, Robert Harkness, an expert on the theoretical supercomputer calculations of radiation from supernovae. Robert is also an expert on airplanes. He knew all about the Quantas stretch 747 that we flew from Los Angeles to Sydney. He had also learned from Brian Warner that Brian had been able to see SN 1987A from the window of the upper-level, first-class lounge for which 747s were so famous.

On the other hand, Robert cannot sleep on airplanes. I can. I had a nap while Robert sat in his seat. I woke up for a meal and then slept again. Robert ate little and sat some more. I awoke feeling great while we were in the middle of our 14-hour flight to Sydney. Although Robert was a bit out of sorts by this time, I asked the flight attendant if we could venture up into the upstairs lounge to try to get a peek at the supernova. She asked the captain and he, in turn, invited us, not into the lounge, but onto the flight deck.

So up we scrambled to meet the crew of relatively young Australians, the pilot Jeff Chandler, the co-pilot, and the navigator. I'm sure this would not have happened on an American airline, and I'm not sure it was strictly legal on Quantas. In any case, the crew were fairly bored from the long flight and keen on the distraction we provided. We asked whether they knew where the Large Magellanic Cloud was. The navigator laughed and replied he had no idea. They flew by computer and never looked at the stars. Robert, no observational astronomer himself, then leaned down and peeked out the window next to Captain Chandler and announced, "There it is!"

Indeed, our flight path was such that the Large Magellanic Cloud was at about 10 o'clock from the nose of the aircraft, easily seen out the captain's left window. It was not trivial to see the supernova. Although it was still fairly bright, it had faded from maximum. My admiration for Oscar Duhalde and what he noted in those first few hours went up. I had brought along some binoculars. With them, I could make out the bright dot of light next to 30 Doradus.

Then Captain Chandler had an idea. He said that fresh oxygen helps visual acuity. He pulled his oxygen mask from its holder. This was not a full-face mask, but tubing that was more reminiscent of the oxygen lines for patients in hospitals. There was a framework that supported the thing over your ears. We spent the next 10 minutes passing around the mask and binoculars. The drill was to take the mask, snort a few deep drafts of oxygen, then rip off the mask (and in my case eye glasses), hold up the binoculars, and peer at the supernova. Frankly, I could not tell that it made any difference, but it sure was amusing! These were not, perhaps, ideal circumstances, but I can say that a few optical photons from the degraded gamma rays from the radioactive decay of supernova-created cobalt made it into my very own retinas. I may die a bitter old man, but it won't be for lack of seeing this remarkable event.

Robert and I spent a couple of days in Sydney among the city lights where viewing the supernova was not practical. We then proceeded to Canberra, site of Mount Stromlo Observatory and the location of the small meeting that was our excuse for this Australian junket. I gave a public talk that first night. I mentioned my curiosity about the native names for the Magellanic Clouds and the next day got a call from a gentleman by the name of Edward Wheeler, no relation that we could identify. He provided me with the names for the Large and Small Magellanic Clouds according to one of the dialects spoken around Sydney when the first British settlers arrived in 1798. The Aborigines speak some 500 languages, so possibilities for other wonderful names like Calgalleon and Gnarrangalleon are enticing. Afterward, there was a clear night, but Robert and I were still exhausted from our trip (and a couple of late nights in Sydney), so we made no attempt to see the supernova that evening. That would have required staying awake until 2 A.M. We had a beer with our host, Mike Dopita, and went to bed.

It clouded up that night. The patch of clouds did not cover all of Australia, but only that fraction we were destined to visit: Canberra, Sydney, and the other major observatory, the Anglo-Australian Observatory at Coonabarabran in the north. By the time we got to Coonabarabran, we were aware that our chances were slipping away. Both Robert and I awoke on the mountain top and watched fog blow over, opening occasional "sucker holes," but never giving a good view of the sky, never mind the Large Magellanic Cloud. We talked a little desperately of getting a car and driving down off the mountain because there was some thinking that the fog might be a localized, mountain-top phenomenon. The bottom line was that we left Australia the next day, having never seen the supernova from the ground. Thank goodness for that Quantas crew.

3. Lessons from the Progenitor

SN 1987A is one of a very few supernovae for which there is any evidence of the star that existed before it exploded. The star was seen in photographs taken for other purposes. It was listed in a catalog of hot stars in the Magellanic Clouds compiled by Norman Sanduleak. The star that exploded was listed by its position in the sky and known as Sk-69 202. You can make it out if you know where to look in Figure 7.4.

Sk-69 202 was not well studied. It was on a list of stars that German astronomer Rolf Kudritzky was investigating intensively, one by one, but it blew up just before Rolf got to it. There is some scientific import to the lack of attention drawn to the star. As Peter Conti, a hot star expert from the University of Colorado, remarked, there was nothing special about Sk-69 202. It did not vary in light output. It did not have any anomalous emission lines. It did not seem to be shedding mass at an especially noticeable rate or in a special way. There was simply no hint at all that Sk-69 202 was special until it disappeared in a violent flash of light. We still do not know why that was so.

A blown-up photographic image of Sk-69 202 is shown in Figure 7.5. The original, larger-scale photo was taken for other reasons, part of a study of star formation in the vicinity of the 30 Doradus nebula,

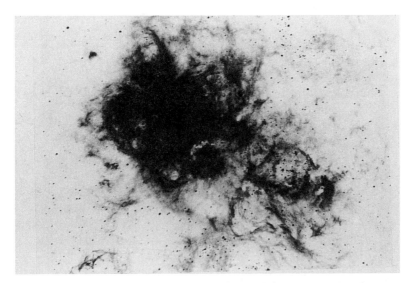

Figure 7.4. Photographic negative of 30 Doradus and Sk-69 202. In this orientation, a small pattern of hot gas on the left somewhat resembles the head of an animal, with two ears, looking to the left. Below the neck of the animal is a single, isolated star. This single (black) dot is Sk-69 202, soon to become SN 1987A. (Photo by You-Hua Chu.)

by You-Hua Chu of the University of Illinois. The big dark patch in the center of Figure 7.5 is Sk-69 202. It is just a point of light, but it looks big because the photographic process smears out the image. The brighter the star, the more intense and the larger the image. This also became known as Star 1, the star that blew up. To the upper right in this image is what is known as Star 2. This is another star in the Large Magellanic Cloud. It is somewhat less massive than Sk-69 202 was. It is not physically or gravitationally close to Sk-69 202, it is several light years away, but it was probably born in the same burst of star formation that gave rise to Sk-69 202 and other fainter stars in this image. Dr. Chu gave me this slide when I went to Champaign-Urbana to present an already-scheduled colloquium on another topic about a week after SN 1987A erupted. She saw something in the photo that was part of a story that played out over the next few months.

When SN 1987A first went off, the vicinity of the supernova shown in Figure 7.5 was lost in the intense glare of the explosion. SN 1987A faded first in the ultraviolet. As it did, Star 2 in Figure 7.5 could be

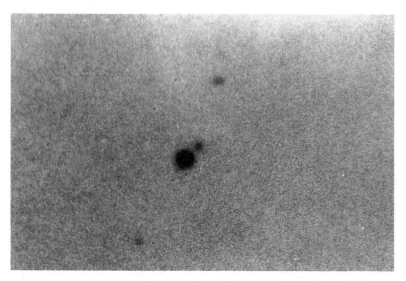

Figure 7.5. Image of Sk-69 202, the progenitor of SN 1987A. Note Star 2 at the upper right, about 2 o'clock, less than one diameter away from the main, dark spot in this negative image. Star 3 is revealed as a slight blurring of the image of Sk-69 202 in the lower left, about 7 o'clock in this orientation. (Photo by You-Hua Chu.)

identified. The surprise was that something was also left behind at the location of Star 1 in the images from the *International Ultraviolet Explorer* satellite, the only ultraviolet instrument available at the time of the explosion. The lingering ultraviolet image left some people wondering whether the wrong progenitor star had been identified. What You-Hua Chu had recognized was that the lower left part of the image in Figure 7.5 was somewhat blurry. She was sure there was a third star there, Star 3, that was obscured by the brighter, smeared image of Sk-69 202 in Figure 7.5. As SN 1987A continued to fade, careful positions were measured, and it was determined that the lingering image was not at the location of Star 1, but slightly offset. There was, indeed, a third star, Star 3. Both Star 2 and Star 3 show up clearly in later images taken with the *Hubble Space Telescope* after the supernova faded (see Figure 7.6). Other people got more credit for resolving this mystery at the time, but there is no question in my mind that Chu knew of the existence of Star 3 within days of the explosion. She scored another coup a decade later at a meeting in Chile to celebrate the tenth anniversary of the discovery of the supernova when she reported that she had discovered the first star to have rings around it like the progenitor of SN 1987A (Section 8).

From preexplosion observations such as Figure 7.5, we know that Sk-69 202 had a mass of about 20 solar masses. This follows from knowing the luminosity. The luminosity is a clue to the mass of the evolved helium core even though that core was buried in a surrounding hydrogen layer. From our knowledge of stellar structure and evolution, we can then estimate the mass that the star originally must have had to make such a massive core. The luminosity suggests that the core was about 6 solar masses, and such a core arises in a main sequence star of about 20 solar masses. The star shed some mass while evolving. The best estimates are that the star retained about 15–18 solar masses by the time it exploded.

Somewhat surprisingly, the star that exploded was not a red supergiant, as might have been expected given the basic theory of stellar evolution and the observation that there are many red giants of 20 solar masses in the Large Magellanic Cloud. Instead, the star was relatively compact and blue. The reasons for this are still not fully understood. The relatively small size produced an unorthodox and somewhat dim light curve. The light curve is by now well understood, given the starting conditions of the star when the explosion erupted. A legion of computer models based on single stars has been calculated in the attempt to understand the compact starting conditions, none of them entirely satisfactory. These models based on single stars may be wrong. The current hot idea is that Sk-69 202 might have been in a binary in which the companion was engulfed in a common envelope and dissolved, leaving only one star to explode. This process might have caused the envelope of the progenitor to contract to a smaller radius and produced some of the other special features of SN 1987A that we will discuss in Section 8. There is no definite sign of any current companion, but that is consistent with none ever having existed, or the companion having been consumed by the supernova progenitor.

4. Neutrinos!

SN 1987A brought us a wealth of new understanding, but the single most important aspect was the burst of neutrinos that were detected from Earth. SN 1987A generated about 10^{57} neutrinos. Most of these went off in directions away from the Earth. Only a tiny fraction arrived at the Earth, and of this number only a tiny fraction interacted with the detectors so that their presence could be recorded. In the case of the neutrinos, the fact that the "observatories" were in the northern hemisphere was irrelevant. The neutrinos, with their ability to interact weakly and hence penetrate matter easily, raced up through the Earth. The same property meant that most of the neutrinos that passed through the detectors also did so without any interaction. Of the original 10^{57}, only nineteen neutrinos interacted with atoms of water in the detectors generating recorded flashes of light. Neutrinos were first detected by the Kamioka experiment in Japan, mentioned in Chapter 1 in the context of the solar neutrinos. Some neutrinos were also seen by a similar experiment in a salt mine near Cleveland and in a special site under a mountain in the Caucasus, what was then the Soviet Union. Those nineteen detected neutrinos were sufficient, however, to show that the basic picture of core collapse was correct. SN 1987A gave birth to extragalactic neutrino astronomy. Unfortunately, with the scant evidence of the nineteen neutrinos, we cannot determine whether the mechanism of the explosion was a core bounce, neutrino heating, or some other related process.

Putting the story together after the fact, astronomers realized that the neutrinos arrived at the Earth before the light. The reason is that the neutrinos are generated in the core collapse, or shortly thereafter, for about 10 seconds. The neutrinos that escape from the newly formed neutron star race outward at very nearly the speed of light. If neutrinos have a small mass as current theories suggest, then they will not travel at quite the speed of light, but so close to it that the difference is negligible. The shock wave that causes the star to explode propagates very rapidly, about one-thirtieth the speed of light. This is faster than the speed of sound in the star, but not at the speed of the departing neutrinos. It took the shock wave about an hour to propagate to the edge of the blue supergiant and generate the first intense burst of light seen by Oscar Duhalde and recorded by Ian Shelton and Rob McNaught. Those first photons were thus a light hour behind the neutrinos, a lag of about 10 million kilometers, about the radius of Jupiter's orbit. The pulse of neutrinos and that first pulse of light raced each other for 150,000 years, but the light could not catch up. The neutrinos arrived an hour ahead of the optical photons. At this moment, more than 10 light years beyond the Earth, the pulse of neutrinos is still ahead of the leading edge of the pulse of light.

5. Neutron Star?

The detection of the neutrinos was dramatic confirmation that a very compact object formed in SN 1987A by the process of core collapse. This result is completely consistent with stellar evolution theory for a star of initial mass about twenty times that of the Sun. The icing on the cake would be the direct detection of the neutron star.

We know that the supernova of 1054 that made the Crab nebula did leave behind a neutron star. This knowledge does not help us to reach general conclusions about how stars explode and make neutron stars because the Crab nebula is peculiar in many respects. It has a large helium content and slower expansion motions than are characteristic of most supernova remnants. Despite the useful observations of the Chinese, we do not know whether it was a Type I of some flavor, a Type II, or perhaps a transition event like SN 1993J. Astronomers of that era could not obtain spectra.

SN 1987A is the best-studied supernova ever, and we know it underwent core collapse, so the potential to learn about neutron star formation is great. As of this writing, however, SN 1987A is 12 years old, and there is still no concrete evidence for a neutron star. This is important because there is just the smallest possibility that the collapse could have generated an explosion and the observed neutrinos, but still ultimately have crushed the nascent neutron star to make a black hole. SN 1987A seems to be a close cousin to the supernova that produced Cas A in about 1680. Both were dimmer than usual, both seem to have occurred in massive stars, and until very recently, neither had obvious evidence for a neutron star. See Chapter 8 for an update on Cas A.

This does not prove that a neutron star is absent in either SN 1987A or Cas A. The neutron star in SN 1987A could be slowly rotating or not very magnetized and therefore not radiating very much. There is also a question of whether the neutron star could be "beaming" its radiation away from Earth as some pulsars are known to do (see Chapter 8). The argument against that is based on the fact that the expanding gas of SN 1987A must surround any pulsar. This gas should absorb any emitted pulsar energy and reemit the energy in all directions. What is certain is that if there is a neutron star in SN 1987A, that 12-year-old neutron star is dimmer, by about a factor of 10, than the nearly 1,000-year-old neutron star in the Crab nebula.

6. The Light Curve

SN 1987A also provided the most direct evidence that radioactive decay of nickel-56 and cobalt-56 can power supernova light curves. Because it was a relatively compact star, Sk-69 202 had to expand farther before it could leak the heat from the original shock. It did not have to expand as far as a Type I, but about ten times farther than a normal Type II exploding in a red-giant state. Thus SN 1987A cooled more than a normal Type II and had less shock heat to radiate by the time it could radiate. This made it dimmer than a normal Type II supernovae. The fact that it was a blue supergiant with a smaller initial radius made SN 1987A naturally dimmer than a normal Type II explosion in a red giant.

Models of the explosion of SN 1987A show that the shock energy dissipated in the expansion about a week after the explosion, yet the supernova did not attain maximum light for 2 months more. That power came from radioactive decay of nickel to cobalt to iron. Models show that the peak light in SN 1987A is produced solely by decay of nickel and cobalt. After the peak light, the light curve declined at a well-defined rate, showing the precise half-life of decay of cobalt-56. From the brightness of the tail, one can read off precisely how much nickel was originally ejected and how much iron will eventually expand into space. The answer is 0.07 solar mass. This is a little on the low side compared to prior expectations but in the range expected for a star of 20 solar mass. In addition, there is direct spectroscopic evidence for the cobalt, and satellites rigged to measure gamma rays detected the gamma rays that were predicted to come from the decay of cobalt. The direct evidence for nickel and cobalt decay in SN 1987A gives us increased confidence that the same process accompanies core collapse in other explosions in massive stars. Understanding these processes in SN 1987A also gives us more confidence to use them in the rather different environment of the thermonuclear explosions of Type Ia supernovae.

7. This Cow's Not Spherical

There is an old joke, one version of which has a scientist hired to study the efficiency of a dairy. He begins his report with the statement, "First we assume all cows are spherically symmetric." This is an in-joke that carries a lot of weight with astronomers. Stars are almost perfectly spherically symmetric because gravity pulls in on them in all directions. Stars are not exactly spherically symmetric, however, if they rotate rapidly or have a strong magnetic field. Still, to make headway in understanding new phenomena, physicists and astronomers have learned that it is often fruitful to make simplifying assumptions to block out the rough truth. Details, out-of-roundness, can be added later as needed. For SN 1987A, it was needed.

The first computer models of SN 1987A assumed that the cow was spherically symmetric. That simplifies the analysis, making minimal computational demands on what are already complex computer calculations. Such simplified models were the obvious place to start. The first clue that they were substantially wrong came from the detection of X-rays. At a meeting in Tokyo (see sidebar) 6 months after the first detection of the explosion, in August of 1987, several theorists presented their predictions that the expansion should lead to the free streaming of X-rays and gamma rays from the radioactive decay in about another year. Japanese astronomers had recently launched a new X-ray satellite. They calmly stood up and reported that they had already detected the X-rays!

The reason for the early onset of X-rays was that SN 1987A was not expanding as a uniform sphere with the hydrogen on the outside, a helium layer deeper in, and the nickel, cobalt, and iron down in the deepest, slowest moving layers. SN 1987A was instead a roiling, turbulent mess that stirred the elements it ejected. Further thought and subsequent computer models showed that fingers of radioactive nickel should, and did, reach out into the outer layers. Streams of hydrogen and helium should plunge inward. The outward mixing of nickel allowed the X-rays and gamma rays to emerge earlier than predicted from the simple models. We learn from our mistakes. By now the understanding of the complicated structure of SN 1987A and how those lessons apply to other types of supernovae has reached a fairly sophisticated level.

8. Other Firsts

Further observations revealed two other "firsts" for SN 1987A. Both were expected at some level, but never before seen. One was the formation of molecules. Molecules of varying complexity fill the interstellar medium. If the density is high enough, single atoms can bind together to form molecules. This apparently happened in SN 1987A.

After about 200 days, SN 1987A showed evidence for at least carbon monoxide (CO) and silicon monoxide (SiO). There are other ways of forming molecules, but one cannot help thinking that the first steps toward molecular complexity that lead to life might begin in supernovae like SN 1987A.

The other interesting observation was to see "dust." The interstellar medium is also full of tiny bits of grit that astronomers call dust. Astronomical dust is interstellar dirt, formed of clumps of graphite (carbon) or sand (silicon oxides) or rust (iron oxides). Theories had predicted that the carbon, oxygen, silicon, and iron in supernovae might in some circumstances coalesce into dust. SN 1987A gave the first firm observational evidence for this process when the light curve got dimmer after about 500 days, as it became shaded in a cloud of its own dust. Studies of this process showed that the dust formed in dense patches, again emphasizing that the ejecta of the supernova were not uniform, but very clumpy.

9. Rings and Things

The most dramatic direct evidence that something about SN 1987A was not sedately spherically symmetric is from the amazing pictures of the rings around the supernova. These were first discovered from the ground but were widely illustrated by images from the *Hubble Space Telescope*. As the epic of SN 1987A unfolded, the *Hubble Telescope* was launched, found to be out of focus, and repaired in a dramatic space walk. The focused *Hubble* images revealed a central ring around the supernova that is tilted in its aspect to us. There are also two fainter rings, nearly but not quite concentric with the first. These preexisting ring structures and the central, expanding supernova ejecta are shown in Figure 7.6. The *Hubble* images also show that the ejected matter is not round in profile, but elongated. This can be seen in Figure 7.7.

The origin of these rings is still debated. They must have formed by matter shed by the progenitor star before it exploded. One popular model is that the star blew a slowly moving wind from its equator while it was a red giant and then a faster wind after it contracted to become the blue supergiant that eventually exploded. The fast wind is supposed to have shaped the slow wind to form the bright ring and to have expanded outward to form the other two rings. Unfortu-

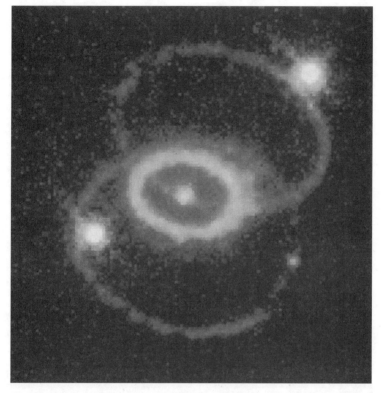

Figure 7.6. The rings of SN 1987A. The matter that forms these rings was shed from the progenitor star before it exploded and was illuminated by the light from the explosion. Note that the rings are not exactly colinear; the edge of the upper ring passes across the central ejecta, whereas the lower ring circumscribes the ejecta and the inner ring. Star 2 in Figure 7.5 is just beyond the upper ring to the upper right. Star 3 is the image to the lower left just inside the lower ring. The smaller bright dot just opposite Star 3 directly on the rim of the lower ring is yet another star in the Large Magellanic Cloud. (*Hubble Space Telescope* photo by NASA.)

nately, computer models show that the inner, bright ring often does not survive the interaction in the form observed. Another hypothesis is that the rings were shed when the progenitor of SN 1987A consumed a smaller mass binary companion.

What has been clear all along is that the inner ring is only a few light years across. The most rapidly moving outer portions of the exploding star are moving at a substantial fraction of the speed of light, at least 10 percent. This implied that in a few years, or perhaps

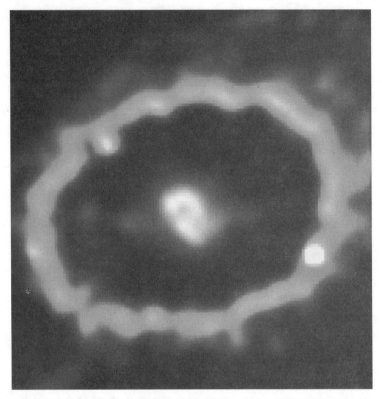

Figure 7.7. The first evidence of the collision of the ejecta of SN 1987A with the inner ring. The spot at about 10 o'clock began to brighten in 1997 as a portion of the fast-moving ejecta (not visible) collided with a clump of matter along the inner edge of the ring. The spot in the lower right (about 4 o'clock) is yet another faint star in the Large Magellanic Cloud, the image of which just happens to fall on the inner ring. Note that the central image of the ejecta is distinctly out of round. It is elongated in a direction roughly perpendicular to the major axis of the inner ring and has a "hole" in it, giving it the appearance of a question mark. Whether the dark spot is a true hole or a patch of obscuring dust is not clear. (*Hubble Space Telescope* photo by NASA.)

a couple of decades, the ejecta should smash into the ring. The result should be a renaissance for SN 1987A. Astronomers expect a new brightening in the optical, the radio, and the X-rays from the gas heated by the collision. The ring is formed of bits and clumps of gas. Each of those should light up when the shock wave hits it, so the ring should sparkle like fireworks over time scales of months to years.

The first estimates of when the collision should occur were based on the notion that there was no material between the supernova and the inner ring to slow the ejecta down. The answer was about 10 years, or, roughly, 1999. More study showed that the space between the supernova and the ring did contain matter. The time for the collision was put off to about 2005. That is not long in the big scheme of things; however, it is long in the life of an astronomer waiting to check a theory.

Given this new time scale, there was thus a little surprise when the *Hubble Space Telescope* revealed that a small portion of the ring had brightened in 1997, as shown in Figure 7.7. Many people thought that the collision had begun. Others worried that there might be some other unexpected anomaly. Ground-based observations from March 1998 showed that many more clumps were lighting up. The collision has begun. Why it occurred faster than the revised estimates is a puzzle. The image of the ejecta in Figure 7.7 resembles a question mark. That is an appropriate metaphor as we continue to follow this piece of astronomical history as it evolves. What is clear is that the next few years will be exciting ones for SN 1987A. This amazing event has much more to teach us.

8

Neutron Stars

Atoms with Attitude

1. History – Theory Leads, for Once

In 1932, the brilliant Russian physicist Lev Landau argued on general grounds that the newly discovered quantum pressure could not support a mass much in excess of 1 solar mass. He addressed his discussion to electrons, but the type of particle did not matter. In 1933, the neutron was discovered, after Landau's paper had been submitted. In retrospect, Landau's arguments applied to the quantum pressure of neutrons as well. An object supported by the quantum pressure of neutrons should be smaller and denser than a white dwarf, but it should have nearly the same maximum mass, about 1 solar mass.

Fritz Zwicky of Caltech was one of the world's first active supernova observers. Quick on the pick-up, Zwicky suggested in 1934 that supernovae result from the energy liberated in forming a neutron star. Not until a year later, in 1935, did the precocious young Indian physicist, Subramanyan Chandrasekhar, present his rigorous derivation of the nature of the quantum pressure and the mass limit to white dwarfs that bears his name.

Robert Oppenheimer made history with his leadership of the Manhattan Project, but among his most widely known papers are two published with students in 1939. The first of these papers used the complete theory of general relativity for the first time to estimate the upper mass limit of neutron stars to be 0.7 solar mass. The second paper explored the result of violating that limit with the resulting production of a black hole. The upper limit to the neutron star is now commonly referred to as the Oppenheimer-Volkoff limit, after the authors. In the 1960s, repulsive nuclear forces between the neutrons were added to the purely quantum effects. As a result, the estimates of the maximum mass of neutron stars rose to between 1.5 and 2.5 solar masses.

In 1964, John Archibald Wheeler suggested that the power radiated by the Crab nebula could plausibly be provided by the rate of loss of rotational energy of a neutron star. This proved to be a prescient guess. At about the same time, Rudolph Minkowski, an old cohort of Fritz Zwicky, was studying the Crab nebula. He pointed out that, although most of the stars seen in a photograph were foreground or background stars, one, apparently buried in the heart of the nebula, had a peculiar spectrum and an abnormally blue color. Minkowski could not prove that this peculiar star was in the nebula. There was not a shred of rational evidence relating Wheeler's speculation to Minkowski's observations, but the relation turned out to be true.

Theoretical astrophysicists often find themselves dragging along behind the observations, trying to explain some exciting new phenomenon ex post facto (quasars represent a superb example). In the case of neutron stars, however, the theorists were way out in front. More than three decades passed from the first theoretical discussions of neutron stars until some confirming evidence came in.

In 1967, Jocelyn Bell was a graduate student working with Anthony Hewish on a peculiar radio telescope at the University of Cambridge in England. The telescope was a series of wires run helter skelter, designed to look for rapid modulation of radio signals by the solar wind. What Ms. Bell noticed among the reams of data was a source of regularly pulsed radio emission. The pulses lasted 0.016 seconds and recurred quite regularly, every 1.33730115 seconds, with astounding accuracy.

The investigators were mystified at first and then, after some contemplation, petrified. There had been a long-standing expectation that any extraterrestrial civilization would signal its existence with some regularly modulated mechanism. The strange signals were dubbed LGMs, short for little green men, and a strong air of secrecy cloaked the lab. This conclusion was too significant to be blabbed about, while further checks ensued.

Soon other such sources were discovered. Significantly, and much to the relief of the researchers, they found the pulse periods were gradually increasing. The fantastically accurate period was not locked in as it would be with an artificial mechanism, but slowly drifted. Whatever these things were, they represented a natural phenomenon. The discovery of *pulsars*, pulsating radio sources, was announced to the world. Anthony Hewish won the Nobel Prize for Physics for the discovery of neutron stars as pulsars in 1974. To the discomfit of some, Jocelyn Bell, whose perspicacity revealed the unexpected signal, did not share in the award. Dr. Bell, a gracious woman, went on to a fruitful career as an X-ray astronomer.

2. The Nature of Pulsars – Not Little Green Men

What were these pulsars? They could not be ordinary stars. Even the light travel time across the Sun is a few seconds, and the pulses in these objects lasted only a fraction of a second. More practically, the fastest motion the Sun could withstand would be if it changed substantially in about a half hour. This is the Sun's dynamical time scale, the time it requires to respond to an imbalance between gravity and pressure. Any global motion of the whole Sun on a faster time scale, whether by rotation, oscillation, or any other mechanism, would mean that the Sun would tear apart.

White dwarfs are more compact and able to withstand rapid movement. One second – a characteristic time between pulsar pulses – is just about the natural time scale for a white dwarf. Just after the discovery of pulsars there was a great flurry of activity exploring white dwarf models for pulsars. The white dwarfs were pictured to be rotating or oscillating. Some people even considered neutron stars. Because neutron stars were even more compact, they would have no trouble responding quickly enough. The natural dynamical time scale for an oscillating neutron star is about 1 millisecond, or 0.001 second, so there was some question why a neutron star should respond as slowly as 1 second. At first, neutron stars were considered a radical, though not impossible, explanation for pulsars.

The studies that showed that the periods of pulsars lengthened with time continued as the theorists thrashed around for a consistent explanation of pulsars. The gradual lengthening of the time between pulses turned out to be a key, if subtle, clue. Studies of oscillating stars show that they tend to respond more rapidly as they lose energy. The reason is that the oscillations themselves tend to make the star somewhat more bloated and unresponsive. As the oscillations die away, the star gets more compact and bounces more quickly. A rough analogy is to drop a ball and listen for the bounces; they become closer together as the ball bounces less and less high. The lengthening of time between pulses suggested that the pulsar phenomenon had nothing to do with oscillations. As a rotating object loses energy, it spins more slowly, and so the time to make one revolution lengthens. This is in accord with the behavior of pulsars, so some rotational phenomenon was considered the most likely explanation for pulsars.

The next major breakthrough came from studies of the Crab nebula. Ten or twenty pulsars had been discovered, all with periods of about 1 second. Then astronomers focused on the strange star Minkowski had pointed out years before. The star turned out to be a pulsar! The period of the pulses was much faster than had been seen in any other pulsar. The time between pulses was only 0.033 seconds. This time is so short that no white dwarf could oscillate or rotate that fast without being torn apart. The pulsar in the Crab nebula had to be a neutron star, and so, presumably, did all the others! Only rotating neutron stars could account for the whole range in periods from fast to slow. A big star cannot rotate rapidly, but a compact star like a neutron star can rotate rapidly or slowly, depending on circumstances.

The pulsar in the Crab nebula rotates relatively rapidly because it was born only a short time ago and has not had time to lose much rotational energy. The pulsars with spin periods of about a second are deduced to be 1 million to 10 million years old. The Crab pulsar is so energetic that it emits pulses of optical light as well as radio radiation.

We still do not understand clearly why the radiation comes from the pulsars in pulses. That radiation comes from the pulsars at all is, however, a clue to another important property. The neutron stars must contain strong magnetic fields to generate radiation. Fundamentally, radiation is caused by wiggling a magnetic field. This causes a wiggling electric field, which in turn causes a wiggling magnetic field, which causes a . . . Coupled wiggling electric and magnetic fields are at the heart of the process of electromagnetic radiation. Without a magnetic field, the rotating neutron star could not emit the kind of radio radiation observed. Thus pulsars must be *rotating, magnetized neutron stars*. That the pulsars are magnetic is not too surprising. Ordinary stars like the Sun generate magnetic fields. If such a star were compressed to the size of a neutron star, the magnetic field would be amplified by a factor of about 10 billion. The resulting magnetic field would be just about what is required to generate the radiation in pulsars. Whether squeezing the field of the star that collapsed to form it is the origin of the magnetic fields of pulsars is still not clear. The newly born neutron stars may act like dynamos and make their own magnetic fields.

The simplest magnetic field a neutron star could have is a so-called dipole field like a bar magnet, with a north pole and a south pole, as shown in Figure 8.1. The lines of magnetic force for such a field are arching loops, out one pole and into the other, exactly like the pattern of iron filings around a bar magnet. If the magnetic field is perfectly aligned with the axis of rotation, there will be no radiation, at least no pulsed radiation. The reason is that the magnetic configuration is too symmetric. If the magnetic field is perfectly aligned, there is no effective change in the magnetic field as the neutron star rotates. A

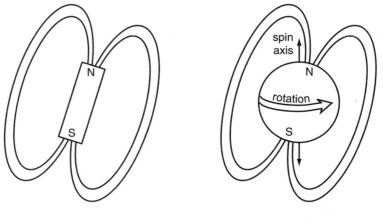

magnet   pulsar

Figure 8.1. The simplest configuration of a magnetic field in a neutron star is a dipole field like a bar magnet with a north pole and a south pole (left). The lines of magnetic force link the poles. To emit radiation, the magnetic axis of the neutron star must be tilted with respect to the rotation axis (right).

wiggling magnetic field is required to generate radiation, and a perfectly aligned magnetic field causes no wiggles as the neutron star rotates.

Radiation will occur if the axis of the magnetic field is tipped with respect to the rotation axis. Then as the neutron star rotates, the magnetic field points in different directions, and the magnetic force at any given point in space varies continuously. This misalignment is not so special a requirement when one considers that the magnetic poles of the Earth are not lined up exactly with the rotation axis and that the magnetic poles even occasionally swap ends.

If pulsar radiation comes from the magnetic poles, we can even understand the pulses because the magnetic poles sweep around like beams from a lighthouse. A pulse would be detected every time a radio "light house beacon" pointed at the Earth. This is the most popular view of the origin of the pulses. Theories have been constructed in which the rotating magnetic fields generate huge electrical fields right at the magnetic poles. The energy in the electric field is so great that it can rip electrons from the neutron star surface or create electron/positron pairs. The particles cause a gigantic spark as they flow along the electric and magnetic fields toward the neutron star or out into space. The spark, like a bolt of lightning at the pole, emits a burst of radio static. This is the particular mechanism envisaged by which the magnetic field "wiggles" and gives rise to radiation.

There is still debate as to exactly where and how this spark forms. As the pulsar rotates around, the magnetic lines of force are carried around with it. Any charged particles caught in the magnetic field are forced to spiral along the field, but they cannot move across the field. The result is that as the neutron star rotates, the particles must rotate as well. All the particles locked to the rotation of the neutron star make a complete circle in the same time, but to accomplish this, the more distant particles, with a greater circumference to travel, are forced to move at tremendous velocities. At not too great a distance from a neutron star, the particles would be whipped around at the speed of light. The path on which particles locked to the neutron star's rotation would move at this limiting speed is known as the *speed of light circle*. The distance would be a thousand miles in the case of the Crab pulsar and 30,000 miles – roughly the Earth's diameter – for a pulsar with a period of 1 second. Because particles cannot move at the speed of light, the particles must be ripped off the magnetic field lines at the speed of light circle. The wrenching process involved would generate radiation. Some theories argue that the great forces generate electron/positron pairs and accelerate them near the speed of light circle so that the "spark" occurs there. Other theories argue that the particles to be accelerated are those pulled from the neutron star so that the spark arises closer to the neutron star surface.

By now some 600 pulsars have been discovered. Most of these are nearby in the Galaxy because their radiation is relatively feeble and cannot be detected from great distances. Extrapolation from the known number of pulsars leads to the estimate that as much as 1 percent of the mass of the Galaxy may be in the form of neutron stars, about one billion of them all told. Most of these would be "dead" pulsars, which could no longer radiate. Pulsars live about 1 million to 10 million years before their magnetic fields decay away or become aligned with the rotation axis, so that no pulses of radiation are possible.

3. Pulsars and Supernovae – A Game of Hide and Seek

When supernovae explode, they inject a large amount of matter and energy into the surrounding gas of the interstellar medium. An explosive "cloud" plows out into the interstellar gas much like a mushroom cloud rises from a hydrogen bomb on the Earth. For a bomb on Earth, the "cloud" rises upward from the ground; for a supernova, the cloud expands outward in all directions. The resulting expanding remnant of a supernova is marked by radiation in the radio that occurs when the shock wave from the supernova compresses and heats the interstellar gas and sends electrons spiraling around the interstellar magnetic field at nearly the speed of light. Interior to the shock wave that marks the point of collision, the shocked gas is so hot it emits X-rays. A *supernova remnant* can span several light years.

These extended supernova remnants live only about 100,000 years before they fade into the general interstellar gas. Pulsars "live" for 1 million to 10 million years. After that time, the neutron star is still around, but it no longer emits radio pulses. Thus pulsars live for about ten times longer than the extended remnants. One expects most pulsars not to be associated with an extended remnant, but that every extended remnant in which a pulsar was born should still surround that pulsar. Most pulsars are not associated with extended supernova remnants, as expected. Strangely enough, the converse is also true.

Most extended remnants show no sign of a pulsar. The Crab nebula is a conspicuous exception to this rule. This negative conclusion has been strongly reinforced by searches for pulsars with X-ray satellites.

This is a puzzling observation. Either no neutron stars are formed in many supernova explosions, or they are not rotating or magnetic so that they cannot emit radio pulses or related traces in the X-ray band, or the pulsars pick up such a high velocity that they escape out of the gaseous remnant. It is possible that in many cases the radio radiation from pulsars is "beamed" so that it does not shine toward the Earth. On the other hand, the X-ray radiation, similar to that emitted strongly from the Crab nebula, shines in all directions so it would be difficult to hide. This raises yet another question. If pulsars are born at the same rate as supernovae explode, but many supernovae do not make pulsars, then apparently there is a way of making pulsars without the associated explosion and optical outburst that identify a supernova. No one knows how this is accomplished, if, indeed, it must be.

This is the context in which one considers the situation with Cas A and SN 1987A. All the evidence is that Cas A represents the explosion of a star of about 20 solar masses. Such a star is predicted to make a neutron star, but until now, none has obviously been seen. The same arguments apply to SN 1987A in a somewhat different context because that supernova is still so young. SN 1987A came from a star of about 20 solar masses. It emitted neutrinos, so we know it had a gravitational collapse, yet any neutron star must be at least ten times dimmer than the 1,000-year-old pulsar in the Crab nebula. Does this mean neutron stars exist in Cas A and SN 1987A but are especially dim? Does this dimness apply to the lack of observed neutron stars in older supernova remnants? Or did Cas A or SN 1987A ultimately create a black hole, and, if so, does this apply to the older supernova remnants? These questions remain central to the study of the final evolution of massive stars.

A new chapter in this story will be written by the *Chandra X-Ray Observatory* launched on July 23, 1999. The very first image obtained by *Chandra* was of Cas A, and to everyone's delight, there was a small dot of X ray emission right in the dead center of the expanding cloud of supernova ejecta. This central source is putting out X-rays with a luminosity of only about one-tenth that of the total light of our Sun. This explains why it was not seen before. It is not clear whether this source is a neutron star or a black hole. The *Chandra* web site asks for readers to vote between these choices. That is an amusing exercise, perhaps, but it is not the way science is done.

This point of X-ray light will be the subject of intense investigation, but a few conclusions are immediately clear. This source is ten thousand times dimmer than the pulsar in the Crab nebula. If it is a neutron star, it is clearly not putting forth the effort to radiate that it might. Even just the heat energy stored in a newly formed neutron star could generate more light than this, never mind any pulsar radiation. On the other hand, a small rate of accretion could make either a neutron star or a black hole shine in X-rays like this, so either could be powered by the fallback of some supernova ejecta that did not quite make it. This discovery also sheds light on the situation with SN 1987A. If a compact object this dim resides in the center of SN 1987A, then it is no wonder that it has not yet been detected. Progress on the study of this point of light in Cas A will undoubtedly also help us to understand whether SN 1987A left behind a neutron star or black hole.

4. Neutron Star Structure – Iron Skin and Superfluid Guts

Neutron stars are sometimes referred to as giant atomic nuclei because, like nuclei, they are composed essentially entirely of baryons, neutrons. Because they are so massive and bound by gravity, neutron stars have a "personality" beyond that of any atomic nucleus.

Neutron stars have about as much mass as the Sun, but, because of their very high densities, they are only 10 to 20 kilometers in radius. Their very outermost layers are of nearly normal composition. There are still protons and electrons. The material is probably mostly iron because all thermonuclear processes should have gone to completion. The topmost material is probably gaseous, an atmosphere hanging above the solid surface, just as on the Earth. One major difference is that in the huge gravitational field of the neutron star, the atmosphere would be only a few meters thick. The solid surface can support mountains and other rugged terrain. Mount Everest dropped onto a neutron star surface would be crushed to a foot or so in height. Typical hills and valleys on the surface of a neutron star would range up to several inches in height.

The outer solid crust of ironlike material on a neutron star would be a few kilometers thick. An important difference in the structure of this material is that the crust is permeated by the huge magnetic field. This magnetic field alters the structure of atoms. Electrons can move along a magnetic field line but cannot move across field lines. This rule applies even to the electrons in atoms if the magnetic field is strong enough. The result is the deformation of atoms into long skinny strings, with the electron clouds elongated along the magnetic field lines and confined in transverse directions. These atoms can in turn be linked to form new kinds of long skinny molecules, which could only exist in the extreme conditions of the crust of a neutron star.

Deeper into the neutron star, electrons are squeezed tightly by the exclusion principle, and the quantum energy they acquire forces them to combine with a proton to form a neutron. The nuclear forces cannot hold a large excess of neutrons into a nucleus, so neutrons begin to leak out of specific nuclei and move around freely in the material. This process is known as *neutron drip*. The densities at which it occurs are higher than the highest density of any white dwarf, but these conditions are still found only a few kilometers deep in the neutron star.

Upon reaching depths where the density is comparable to the density of normal atomic nuclei, nothing resembling a normal atom can exist. The material is essentially all neutrons, although there is a scattering of protons and electrons. The few electrons can still exist because they are so sparsely spread that the effects of exclusion are small, and their quantum energy is not appreciable. There is one proton for every surviving electron to balance charge. The densities are so high that the exclusion effect on the neutrons is dominant and their quantum energy, moderated by effects of nuclear forces, provides the pressure to support the neutron star. The quantum uncertainty in the "cloud" that represents a massive neutron is smaller than that for the cloud of the smaller-mass electron. This is why electrons feel squeezed first, and neutrons must be raised to much higher densities before the exclusion of one neutron by another has an appreciable effect.

A remarkable transition in the nature of neutron-star material is made at higher densities. The nuclear forces between neutrons have another important role besides just altering the pressure. The nuclear forces cause the quantum waves that represent the neutrons to line up in a special way that minimizes the repulsive nuclear forces. The result is that the neutrons are thought to form what is called a *superfluid*. A superfluid is a special state of matter in which all the particles flow in consonance and the result is absolutely zero viscosity, no resistance to motion. Water has much less viscosity than molasses, but a superfluid has none at all! Physicists have created superfluids in the laboratory by cooling liquid helium to near absolute zero. This reduces the thermal energy in the helium, and helium has no interfering chemical reactions because it is a noble gas. The result is that the quantum properties dominate, and the quantum waves of the helium atoms can line up in such a way as to form a superfluid. The resulting material flows so easily that if care is not taken, it will flow up the side of the beaker and out of the experiment! Lev Landau, with whom we introduced this chapter, won the Nobel Prize in Physics for his work on liquid helium in 1962.

At the highest densities in the center of a massive neutron star, the quantum effects among the neutrons can cause yet another arrangement of the structure. Theories predict that the neutrons will clump together into a rocklike solid. This material would be somewhat akin to the solid crust. In the crust the solidification is due to electrical forces on the electrons, whereas in the core the solidification is due to nuclear forces among the neutrons. At these most extreme densities, the huge gravitational energy can be converted into mass. Exotic particles that do not normally exist in nature could spring spontaneously into existence, but there is no proof that such processes occur.

This picture of the interior of a neutron star just sketched follows from the theoretical extrapolation of known physics to extreme conditions. Fortunately, there is some evidence that the picture is at least qualitatively correct. This evidence comes from "glitches" observed in the rate of pulses from pulsars. As we have said, pulsars generally slow down with time in the sense that their pulses slowly get farther and farther apart. This effect is quite gradual, of order of one part in a million per year, and it is only due to the exceedingly accurate rate of pulses that the slowdown can even be detected. Occasionally, however, a pulsar will speed up for a short while, and the time between pulses will become shorter. After some time, the pulses will settle back into their old pattern of gradual slowing. This behavior is known as a "glitch," which means, in general, an unexpected interruption or change in behavior. Glitches have been observed in a few of the youngest pulsars. Apparently, the older pulsars have settled down into a state where they do not glitch anymore. The Crab pulsar has been observed to glitch. There is another supernova remnant in the direction of the constellation of Vela. This supernova remnant is only about 10,000 years old. It also contains a pulsar that has been observed to glitch.

No one has seen a pulsar in the process of glitching. Rather, the pulsar is observed at one time and then a little later, and the period is found to be slightly shorter. From such observations a few days apart, one can conclude that the glitches happen on a time that is shorter than a few days (possibly much shorter), but no more accurate statement can be made. The thing that is of particular interest is that after a glitch, the pulsar requires a considerable time, of order a month, to return to its original period and resume the same gradual lengthening of the period. That the time to return to normalcy is so long seems to strongly suggest that the inner portions of the neutron star are superfluid.

Glitches are thought to occur as a neutron star adjusts itself to the loss of rotational energy as it slows down. The understanding of how that adjustment occurs has evolved over the decades since glitches were discovered. An early model envisaged the spinning neutron star to form an equatorial "bulge" that was frozen in when the neutron star cooled and its outer layers solidified. As the neutron star spun more slowly, the bulge would settle by cracking and breaking. Conservation of angular momentum would cause the neutron star crust to rotate slightly more rapidly when the crust broke and settled into a smaller radius. This was thought to represent the formation of the glitch. The slow healing time was then thought to represent the long time necessary for the outer solid crust to bring the inner, zero viscosity, superfluid core into a common spin rate, after which the whole neutron star would begin to lose rotational energy and once again begin to spin ever more slowly. The reason to mention this picture is that it is a simple physical one that was reasonably easy to describe in lectures. I used it for decades, and it appears in other books. It is also wrong. More careful study showed that the mechanism of glitches is more interesting and subtle. The idea of crust cracking has survived in another context that will be described in Section 10.

The current model for glitches is based on considerations of exactly how the magnetic field that is such an obvious part of the external aspects of a pulsar threads the inner superfluid core. It turns out that a magnetic field cannot penetrate the superfluid, but only normal matter. For the magnetic field to thread the superfluid core, there must be "vortices" of normal matter that extend through the superfluid core, roughly parallel to the spin axis of the neutron star. The spinning vortices of normal matter are the repository of the angular momentum of the material in the inner core. The vortices of normal matter also provide the path for the magnetic field to pass from the north to the south pole within the neutron star. The vortices that allow normal matter and magnetic field to thread the superfluid are "pinned" to irregularities in the normal matter of the outer crust. In this picture, a glitch occurs because the vortices have a memory of the past when the outer crust was spinning faster. At intervals, some of the vortices unpin from the crust and coalesce, allowing the whole neutron star to adjust to its slower rotating, lower angular momentum state. Although the whole neutron star adjusts to the lower rotational state, this unpinning causes the outer crust to temporarily rotate more rapidly, giving rise to the glitch. As the neutron star attains its new equilibrium rotational state, the vortices again pin to the crust and slow it down so that the gradual slowing of the whole neutron star can continue. The bottom line is still that the glitch phenomenon cannot be explained without invoking a superfluid core.

5. Binary Pulsars – Tango por Dos

The accurate periods of pulsars make excellent clocks. If the clock were to move, the *frequency* of the pulses would be changed by the *Doppler shift*. The frequency of the radio emission would also be changed, but the radio radiation is continuum radiation, which, without spectral "lines," specific identifiable frequencies, gives no detectable Doppler shift. The pulses themselves are a marvelous substitute. With this clock, astronomers can look for periodic changes in the velocity of a pulsar that would indicate that the neutron star was in orbit. The evidence shows that to a high degree of accuracy the vast majority of pulsars are not in binary star orbits. Astronomers were very excited when in 1975 careful searches paid off, and a radio pulsar was discovered to be in a binary orbit. Since then, eleven more binary radio pulsars have been discovered. They are the exception that proves the rule: the vast majority of the known pulsars are single stars.

The discovery of the first binary pulsar led to a host of interesting results. The orbit was worked out from the Doppler shift of the pulsar period, and the prediction was made that any companion star of ordinary size would cause the eclipse of the neutron star once each orbit. No eclipse was seen. The lack of an eclipse implies that the companion star is itself a compact star, probably a white dwarf or neutron star.

Nature has been kind to put neutron stars in binary orbits. Study of the binary orbits allows the determination of the neutron star masses, a fundamental property that cannot be accurately measured by any present techniques for the multitude of single radio pulsars. The period of the orbit gives information about the masses of the stars, using Kepler's third law. The mass of the first binary pulsar is one of the few known neutron star masses. Both stars seem to have a mass of very nearly 1.4 solar masses. Other binary neutron stars have also had their masses weighed, and they also appear to have very nearly this mass. The coincidence of this number with the Chandrasekhar limit requires some comment. If a white dwarf attained the Chandrasekhar limit and collapsed to form a neutron star, the neutron star would be somewhat lower in mass. This is because some energy is inevitably ejected in the process of forming the neutron star, if only in the form of neutrinos. A great deal of energy must be ejected and the mass equivalent, in terms of $E = mc^2$, of the minimum energy loss is about 0.2 solar mass. To make a neutron star of 1.4 solar mass, the initially collapsing object would have to be 10 or 20 percent more massive, and hence somewhat greater than the Chandrasekhar mass. Just why neutron stars should form from cores of a precise mass that somewhat exceeds the Chandrasekhar mass is not clear.

The accurate orbital timing of the first binary pulsar showed that the orbit was decaying. The two stars are slowly spiraling together. Recall the final evolution of two white dwarfs from Chapter 5. They are imagined to spiral together as they give off gravitational radiation. In the binary pulsar system, the change in the orbit is precisely what would be predicted as the result of gravitational radiation. With one stroke, this observation confirms, indirectly but strongly, the predicted existence of gravitational radiation by Einstein's general theory and shows that gravitational radiation works in binary systems to draw stars together, just as the astrophysicists had predicted. What ever the companion of this binary pulsar, white dwarf, or neutron star, gravitational radiation will eventually cause them to collide and merge. The discovery and analysis of the binary pulsar and the remarkable proof of gravitational radiation led to the award of the

Nobel Prize to Joe Taylor and Russell Hulse, the radio astronomers at the University of Massachusetts (now both at Princeton) who made the discovery and analysis of the first binary pulsar. For this second Nobel Prize for work on neutron stars, the important contribution of the graduate student (Dr. Hulse) was recognized.

The binary pulsars, by being the exception to the rule, also lead us to ask why the strong majority of pulsars are not in binary systems. The binary pulsars provide a clue to the answer. One possibility is that neutron stars are commonly ejected from binary orbits by the explosion that creates them. Arguments based on conservation of energy and angular momentum show that if half the total mass of a binary system is ejected in an explosion, the system will be disrupted with the two stars flying off in opposite directions. In addition, pulsars are observed to sail through space at rather high velocities. There are a number of reasons to think that pulsars are given a "kick" by the process of violent gravitational collapse that creates them. Such kicks will also help to tear neutron stars away from any binary companion. Ejecting matter in the explosion and kicking the pulsars probably account for most of the single pulsars. The exceptions can also be understood at some level. For one thing, the star that blows up will frequently be the less massive star because it will have transferred mass to the companion. If the exploding star contains less than half the total mass of the two stars combined, then it cannot eject more than half the total mass, and the binary system will not be disrupted. The kicks to newly formed neutron stars may not be delivered in random directions, but inasmuch as they are, some of the kicks could help to keep the neutron star in orbit despite the loss of mass and gravity from the binary system by the supernova process itself.

The circumstantial evidence that Types Ib and Ic supernovae arise from massive stars that have lost their outer envelopes by mass transfer suggests that they create neutron stars in binary systems. Whether these neutron stars remain in the binary is not clear. There is a strong suspicion that, for systems in which the neutron star is still in a binary, the neutron star was born in some version of a Type Ib or Type Ic supernova explosion.

There may be another reason why the radio pulsars, in particular, are mostly single. An important feature of the first binary pulsar is that the companion star is known to be compact. No mass is being transferred in the system. As we will see in the next section, neutron stars are known to exist in binary systems for which the neutron star is not a radio pulsar. These systems are transferring mass. One reasonable hypothesis is that mass transfer prevents the emission of radio pulses by blocking the radio emission or by shorting out the sparking mechanism and preventing the radio radiation in the first place. With this picture, one would say that the binary pulsar is special not because the neutron star remained bound in a binary system, but because the companion star is unable to transfer mass and spoil the radio pulses. Those neutron stars that were always single stars or that were ejected from binary systems have no problem because they have no companion to interfere. Most neutron stars left in binary systems are not radio pulsars because they have the misfortune to be neighbors to a living star that insists on sharing some of its matter.

6. X-Rays from Neutron Stars – Hints of a Violent Universe

X-ray observations have been mentioned where appropriate throughout this book. The next subject owes its very existence to the advent of X-ray astronomy, however, and so a word of history is in order. In the last three decades, the science of X-ray astronomy has matured to become a major independent branch of astronomy. X-rays must be collected above the absorbing shield of the Earth's atmosphere. The first observations were made with brief rocket flights that only tantalized the scientists that launched them. There were glimpses of intense sources of high-energy X-rays.

The revolution in X-ray astronomy began with the launch of a small astronomical satellite dedicated to the detection of X-rays in 1972. The satellite was launched from a site in Kenya and was called *Uhuru*, the Swahili word for freedom. This first satellite could not locate the source of any X-ray emission very accurately, and, although better than rockets, it was not tremendously sensitive. *Uhuru* was on station for a long time compared to a rocket at perigee, however, and it could look for X-rays for orbit after orbit. The result was stupendous. The whole sky was alight with X-rays. It was like Galileo's invention of the telescope: to look with a new tool and to find that previously unknown or inconspicuous objects glared forth when examined properly. X-rays were seen from stars, from galaxies, from every direction! Above the protective layer of the atmosphere, the Universe was a far more violent place than astronomers had suspected.

Many X-ray satellites have been flown in the last 30 years. Several have been launched by the United States, others by European countries. Japan has had a very successful series of satellites and nearly took over the field when the U.S. support for X-ray astronomy lagged in the 1980s. Russia has also had a number of successful experiments. A major step of this first burst of activity in a new field was the launching by NASA of a large satellite in 1978 bearing the name *Einstein*, because it was the centennial year of his birth. This satellite contained a device that could focus X-rays like a proper telescope. It could measure details in an X-ray picture with an accuracy of 1 arc second, equivalent to that of ground-based optical telescopes. In 6 years, the science of X-ray astronomy made an advance in sensitivity and detail equivalent to the leap from Galileo's first telescope to the giant modern reflectors. The new *Chandra Observatory* mentioned in Chapter 1 is the latest step in this progression, and there are more and better projects under construction and on the drawing boards.

One of the subjects to benefit most greatly from the new science of X-ray astronomy was the study of neutron stars. This is because the great gravity of these objects causes tremendous heating of any matter that falls upon them. The matter becomes so hot that the maximum intensity of radiation comes in the X-ray portion of the spectrum. Under proper circumstances, neutron stars are just natural X-ray emitters.

Some of the first X-ray sources examined with *Uhuru* showed a peculiar behavior. The intensity of the X-rays was not constant, but faded away at regular intervals, typically every few days. Most of the scientists who worked on the early X-ray experiments building the detectors were physicists, not astronomers. The erratic behavior in the signal puzzled them. Astronomers – at least many amateurs who delight in such things, if not the professionals who specialized elsewhere – would have immediately identified the cause. The problem was that the X-rays were being eclipsed. The X-ray source was in a binary star orbit and was simply disappearing behind the other normal star periodically. This companion star was the source of matter that fell onto the neutron star and produced the X-rays.

This understanding led to a rapid series of identifications of orbiting neutron stars. A major new branch of astronomy was born almost overnight as the new sources were identified and characterized, and theorists rushed to understand their properties. The X-ray observations provided an exciting new way to probe the nature of mass core collapse to form a neutron star and an explosion. Betelgeuse is nearly a canonical candidate for a Type II supernova explosion. We do not know exactly when it will explode. The final stages after a star of this mass becomes an extended red giant are typically no more than 10,000 years. We do not know when in the next 10,000 years it will explode (it may be tomorrow!), but we can estimate the progression of events when it does.

Upon core collapse, Betelgeuse will emit 10^{53} ergs of neutrinos, each with an energy characteristic of a nuclear reaction. This burst of neutrinos will take about an hour to pass through the hydrogen envelope and into space. They will arrive in the Solar System 427 years later and be the first indication that Betelgeuse has erupted. These neutrinos will deliver about 2 $\times$ 10^8 recoils in the body of a 100-pound woman. This effective level of radiation exposure is far less than a lethal dose (by a factor in excess of 1,000, depending on how the energy is actually deposited) but might cause some chromosomal damage. The shock wave generated by the collapsing core and the formation of a neutron star will require about a day to reach the surface. The breakout of that shock will generate a flash of ultraviolet light for about an hour that will be about 100 billion times brighter than the total luminosity of the Sun. This burst may not exceed the ultraviolet light from the Sun at the Earth, but could affect life on outer satellites, if there is any, or any explorers from Earth, if we have ventured far from the Sun by the time this happens. This blast of ultraviolet might cause some disruption of atmospheric chemistry. The ejecta of the supernova will expand and cool after shock breakout, and the total luminosity will first dim and then rise to maximum in about 2 weeks as the supernova material expands to about 100 times the Earth's orbit and the photon diffusion time through the expanding matter becomes comparable to the time required for appreciable expansion of the matter. The total luminosity will then be about a billion times that of the Sun. At its distance, Betelgeuse will be a factor of about one million dimmer than the Sun, magnitude $-$ 12, about the same as a quarter Moon. This phase will last during the "plateau" phase of the light curve, 2 or 3 months. The observed surface of the supernova during this interval will be roughly constant at an effective temperature of about 6,000 K, slightly hotter than the Sun. After the hydrogen envelope has expanded and electrons and protons have all recombined to make neutral hydrogen atoms, the envelope will be nearly transparent, and the light curve will begin a rapid decline.

In a typical supernova of this type, the emission is dominated for the next year or so by radioactive decay of cobalt to iron (nickel will have already decayed away). The expanding envelope of hydrogen is likely to remain opaque to these gamma rays until substantial decay has occurred, so such an event is unlikely to provide a substantial source of gamma rays. If transfer, accretion disks, and the structure and behavior of the neutron stars themselves. Although the existence of accretion disks had been demonstrated in the cataclysmic variables, it was the exciting new realm of neutron star X-ray sources that resulted in the sudden growth of interest and developments in the understanding of accretion disks.

Over the next few years after the launch of *Uhuru*, X-ray astronomers realized that there were two basic classes of binary neutron star X-ray sources (and a handful of oddballs that resist categorization). The first class consists of a neutron star in orbit about a normal, fairly low-mass star. The other class consists of neutron stars in orbit around high-mass normal stars. In this case, the normally evolving star typically has a mass in excess of 10 solar masses.

The classic example of the first type is the first X-ray source discovered by *Uhuru* in the direction of the constellation Hercules, the system named Hercules X-1. Detailed studies over decades have shown that Her X-1 is a nearly textbook example of mass transfer to a neutron star in a binary system, as shown schematically in Figure 8.2. A star of about 2 solar masses, slightly evolved on the main sequence, is filling its Roche lobe and transferring mass. The mass settles into an accretion disk. As friction operates in the disk, the matter spirals down toward the neutron star and gets heated. In the inner portions of the accretion disk, the orbital velocities are very high, so the frictional heating is strong, and the material in the disk itself emits X-rays. As the spiraling matter gets near the neutron star, the magnetic field of the neutron star channels the matter toward the magnetic poles. As the material finally lands on the surface of the neutron star, the impact causes more heating and further X-rays.

Although X-ray satellites are crucial to the discovery of X-ray sources, one should not forget that the astronomy advances most efficiently where standard earthbound optical techniques can be brought to bear in complementary studies. This is because, as a matter of practice, there is a tremendous amount of information available in the photons emitted in the optical band. This is, after all, where most stars emit the majority of their radiation. Most of our practical knowledge of the Universe is obtained in the optical, so X-ray (or radio, infrared, or ultraviolet, or gamma ray) information must be integrated into the realm of classical optical astronomy to come to full fruition.

As an example, studies of Her X-1 would be woefully incomplete

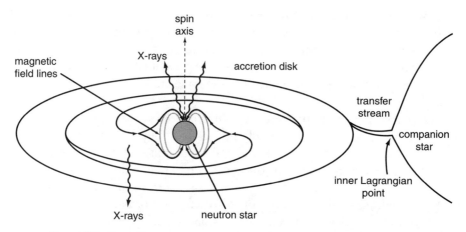

Figure 8.2. Binary X-ray sources consisting of a neutron star with a low-mass companion, like Hercules X-1, are very similar to cataclysmic variables, but with the white dwarf replaced with a neutron star. The companion star, often a main sequence star, transfers mass from its Roche lobe through a transfer stream that collides with an accretion disk. The matter joins the disk and spirals slowly down toward the neutron star. When the magnetic force of the neutron star exceeds the pressure forces in the disk, the matter is diverted to follow lines of constant magnetic force. These paths lead to the magnetic poles of the neutron star. X-rays can be emitted from the inner, hot portions of the accretion disk and from the magnetic poles where matter actually strikes the neutron star surface.

without the optical studies of the companion star. It is the optical studies that tell us the type of star, its evolutionary state, and the fact that it is filling its Roche lobe. Coupled optical and X-ray studies were used to completely characterize the orbits of the two stars and to obtain a direct measure of their masses using Kepler's law. The mass of the neutron star comes out to be very nearly 1 solar mass. This mass seems to be significantly less than the 1.4 solar masses that has been measured so precisely for several of the binary pulsars, as mentioned in Section 5. There is no understanding of why this should be so. It is presumably an accident of birth of an especially low-mass progenitor core, but it might have involved an especially large ejection of the mass from the collapsing core. In this game, even "typical" objects are not so typical.

The observations of Her X-1 suggest that a star of initial mass between 10 and 15 solar masses evolved and shed its envelope. The bare core probably evolved on its own for a while and then collapsed. Like cataclysmic variables, there is a strong hint in Her X-1 that the original evolution was not just a simple case of one star losing mass to the other. For one thing, the two stars are too close together now for the first star to have developed a dense core and red-giant envelope. Also, the relatively low mass of the companion star suggests that it did not accept all the mass that the first star lost. Her X-1 is probably another example of common-envelope evolution in which the 2-solar-mass star was engulfed in the envelope of the more massive star. Much of the first star's envelope was presumably lost out of the system, and the core of the massive star and the smaller-mass companion spiraled together. Perhaps the smaller star filled its Roche lobe while still enshrouded in the envelope of the other. Whether any of this helps to explain the relatively low mass of the neutron star is not clear.

The other kind of binary X-ray source systems, those with high-mass normal companions, is typified by the third X-ray source *Uhuru* discovered in the direction of the constellation of Centaurus, Centaurus X-3. The basic difference between Her X-1 and Cen X-3 is that the mass-losing star in the latter is fairly massive, about 20 solar masses. This turns out to make an important modification to the mass transfer process, if not the ultimate outcome. When Cen X-3 was first discovered and the companion optical star identified, attempts were made to work out the orbits. According to the standard picture, the assumption was made that the companion filled its Roche lobe in order to transfer mass to the neutron star. The answers that emerged did not make sense. The mass of the neutron star was derived to be so low, about 0.1 solar mass, that the gravity should be so weak that any neutron star should expand to be a white dwarf instead.

The problem was that the companion star does not fill its Roche lobe! Rather, such a massive star blows an appreciable stellar wind. It loses mass through this wind whether it has a companion star or not. In this case, however, there is a neutron star the gravity of which reaches out and ensnares some of the passing wind. The matter from the wind then settles into an accretion disk. With this picture, things make more sense. The orbital information from Cen X-3 is not as accurate as that from Her X-1, never mind the binary pulsars. The best estimate for the mass of the neutron star comes out to be a little more than a solar mass, but a mass of 1.5 solar masses cannot be excluded. This is a reasonable result.

The disproportionate mass between the neutron star and the massive normal companion in Cen X-3 has one interesting consequence. The neutron star raises tides on the surface of the companion, just as the Moon does on the Earth. Energy is expended in dragging those tides around, and the energy comes out of the orbit, causing the neutron star to spiral toward the other star. If the companion is not too massive, the tidal drag causes it to spin faster until the companion rotates at exactly the speed that the neutron star orbits. Then the tide just sits in one place on the surface of the star, and there is no drag. For a massive companion, however, there is too much inertia. The central star and the tides always lag behind the orbital motion, dragging the neutron star down. There is no limit to this process, and eventually the neutron star should collide with and disappear into the companion star. The neutron star could spiral to the center, swallow matter from the star, collapse to make a black hole, and then eat the whole star! This may be the fate in store for Cen X-3.

Her X-1 and Cen X-3 share another very important feature. The X-rays they emit come in pulses, 1.2 seconds apart for Her X-1 and 4.8 seconds for Cen X-3. The behavior is very reminiscent of the pulses from radio pulsars, but the energy is coming in the X-ray portion of the spectrum. In addition, for extended periods of time the pulses get steadily more rapid, whereas, except for glitches, the radio pulses slow down.

Despite the exotic nature of the radiation, the X-ray pulses are easier to explain than the radio pulses. Much of the explanation borrows heavily from the knowledge gained by studying radio pulsars. The neutron stars are presumed to be magnetized and rotating. The crucial difference is that, whereas a pulsar must generate radio radiation by its own devices, the X-rays are caused by an external agent, the dumping of mass upon the neutron star.

With the presence of the magnetic field, the matter arrives at the neutron star in a special way that promotes pulses. The matter spirals down in the accretion disk until it encounters the outer reaches of the magnetic field. At that point, the matter finds that it cannot continue in orbit because it cannot move across the lines of magnetic force. Rather, the matter falls along the lines of force, as shown in Figure 8.2. These lead naturally to the north and south magnetic poles of the neutron star. The matter is channeled so that it falls selectively on the magnetic poles, not at random on the surface of the neutron star. The intense X-radiation then comes from the magnetic poles, as if there were two bright spots on an otherwise dark surface. If the magnetic axis is misaligned with the axis of rotation, then, as the neutron star spins around, first one then the other bright spot points at the Earth, just like a lighthouse. The observer detects a pulse of X-rays as the pole is swept into view by the rotation. With mass transfer, one can understand fairly easily why the radiation comes from the poles and hence why there are pulses.

The influence of mass transfer also explains why the pulses tend to speed up rather than slow down. There are two competing effects. The loss of energy in the radiation tries to slow the neutron star down. The matter arriving from the accretion disk, however, carries with it the angular momentum of its orbit. As the matter lands on the neutron star, the spin is transferred to the neutron star. This turns out to be the dominant effect in many circumstances, and the neutron star rotates faster and faster until the mass transfer stops or the neutron star is rotating as fast as the accreting matter where it begins to interact with the magnetic field. If the neutron star tries to rotate too fast, its magnetic field acts like a paddle to splash matter out of the accretion disk, which slows the neutron star down. Both Her X-1 and Cen X-3 have gone through episodes lasting a couple of years where they have stopped speeding up (Cen X-3) or have even tended to spin more slowly (Her X-1). This is presumably because they have ejected matter or the rate of mass transfer has declined so the accretion disk has retreated, allowing the neutron star rotation to slow. Even though the spin-up by accretion makes good sense, the slow-down process must be rather prevalent because many X-ray pulsars have rather long periods, some as long as 800 seconds.

7. X-Ray Flares – A Story Retold

Recall from Chapter 5 that there were two basic classes of flaring binary white dwarf systems: the dwarf novae where the accretion disk is the source of the activity and classical novae caused by thermonuclear explosions on the surface of the white dwarf. Suppose the white dwarf were replaced by a neutron star. Similar phenomena will occur.

X-ray astronomers see several accreting neutron stars in the Galaxy that are labeled as *X-ray transients*. In this context, the general word "transient" refers to a particular phenomenology, implying a particular physical cause. Every few years, these X-ray transients emit a flare of X-rays that lasts for about a month or so. At least two of these systems are well studied and are known to be in binary systems. There is a strong suspicion that the process causing this outburst is similar to that in dwarf novae, an instability in the flow in the accretion disk. The accretion disk instability described in Chapter 4 does not depend sensitively on the nature of the object around which the disk circles. If matter flows into the disk from a companion star at an appropriate rate, the disk will go into the storing and flushing mode that characterizes the dwarf novae. If the object receiving the mass is a neutron star, however, then in the flushing phase, matter from the disk is spiraling down onto a neutron star. The matter gets intensely hot and emits X-rays. The time scales are somewhat longer in the X-ray transients than in dwarf novae, and there are no quantitative models, but the disk instability is a plausible picture for the origin of the X-ray transients.

There is also a neutron star analog of classical novae. In 1978, a fascinating new class of X-ray sources was discovered. Russian scientists first noticed the phenomena. Some X-ray sources show an occasional brief, strong burst. The power rises in about a second and then decays over the course of the next minute or so. The bursts recur every few hours more or less randomly. After the Russians reported these bursts, a search of old *Uhuru* data also showed the effect. The American astronomers just had not noticed it at first in the welter of data with which they had to deal.

The display in the X-ray bursts is not like the rather demure pulses from Her X-1 and Cen X-3 or like the occasional flares of the X-ray transients. The bursts are very energetic compared to the pulses of Her X-1 or Cen X-3. They are comparable in power to the X-ray transients but much shorter in duration. They call for a completely different physical explanation.

Of the more than 100 X-ray sources in the Galaxy with low-mass companions, about 40 are *X-ray bursters*. None of the binary X-ray sources with high-mass companions display this behavior, and neither do the few low-mass systems that display X-ray pulses like Her X-1. Like the general population with low-mass companions, the X-ray bursters tend to cluster toward the center of the Galaxy, as do the oldest stars in the Galaxy. At least nine of the X-ray bursters are seen to be in globular clusters that are also old assemblages of stars. Most X-ray bursters show no evidence for binary motion, but evidence has been reported for orbital motion in at least one X-ray burst source. The guess is that all these systems are in binary systems, but nature conspires to hide the fact. If the systems are seen edge-on, it is most easy to determine the Doppler motion due to their orbit, but in this case the neutron star and its X-rays can be obscured by the accretion disk. If the system is nearly face-on, the X-rays can be seen, but the orbit is difficult to determine because all the motion is almost at right angles to the observer. The Doppler shift only registers the component of motion directly toward or away from the observer. The X-ray bursters do not show any sign of X-ray pulses (an exception will be described later). The interpretation is that the neutron stars in these systems have very low magnetic fields, so matter is not focused on the magnetic poles, and there is no X-ray "lighthouse" effect.

The theory for the burst sources is based on thermonuclear explosions on the surface of the neutron stars. Calculations have shown that as hydrogen accretes onto the surface of a neutron star, it is heated and burns in a regulated fashion. Under proper circumstances, the resulting helium, however, piles up in a layer supported by the quantum pressure. As we have seen in several instances, this condition leads to unstable burning when the helium finally gets hot and dense enough to ignite. The X-ray bursts are thus thermonuclear explosions on the surfaces of the neutron stars. There is therefore a direct parallel for this explanation for the X-ray bursts and the explanation for the outbursts in the classical novae, the basic differences being in the nature of the compact object doing the accreting. Because of the high gravity of neutron stars, relatively little, if any, matter is ejected from the neutron star in an X-ray burst. The high gravity also causes the very short time scale of the explosion on the surface of the neutron star as compared to the effects in a classical nova that can linger for a year or more.

The theory of these nuclear outbursts shows that they only occur if the rate of accretion of matter onto the neutron star is relatively sedate. This allows the layer of helium to build up supported by the quantum pressure. At high accretion rates, the helium stays hot, is supported by the thermal pressure, and burns in a regulated, nonflaring way. One of the implications of this theory is that if the neutron star is strongly magnetic, then even a sedate rate of accretion will be focused onto the magnetic poles, giving an effective high rate of accretion at those two spots. That will give the circumstances for hot magnetic poles and X-ray pulses, but it will mean that the rate of the accretion at the poles is high enough that the helium will ignite and burn in a regulated way. This is another reason to argue that neutron stars that show X-ray pulses have large magnetic fields and no X-ray bursts and neutron stars that show X-ray bursts have small magnetic fields and do not display X-ray pulses.

The Eddington limit discussed in Chapter 2 plays an interesting role in the neutron star accretion process associated with these X-ray burst sources. Recall that the Eddington limit is a limit to how bright an object can be without blowing away matter by the sheer pressure of the outflowing radiation. The Eddington limit depends on the gravity of the object, and so the limiting luminosity scales with the mass. For accreting neutron stars, there is a close coupling between the mass and the luminosity because the luminosity is caused by the infalling matter. This means that if the matter falls in at too high a rate, intense radiation will be generated. The infalling matter will be blown away rather than accreting. If too much of the infalling matter is blown away, however, then there is not enough radiation to blow the matter away, and the infall can take place. The result can be to balance things so that some matter is blown away and some accretes. The luminosity adjusts so that the Eddington limit is not violated. Many of the binary neutron star X-ray sources have luminosities somewhat below the Eddington limit, as if they had made their accommodation with the limiting luminosity. In the observed X-ray bursts, the luminosity rises until it bumps right into the ceiling of the expected Eddington limit for an object of the mass of a neutron star, about 1 solar mass.

At least one binary neutron star system, Centaurus X-4, displays both X-ray transient outbursts and X-ray bursts. As the X-ray flux from Centaurus X-4 declined from one month-long flare of the X-ray transient variety, it showed another brief flare of the X-ray burst variety before proceeding to decline. Presumably an accretion disk instability flushed matter down toward the neutron star creating the X-ray transient. As matter accumulated on the neutron star, it underwent a thermonuclear outburst. Then the disk went into its storage mode; there was no fresh mass added to the neutron star, so no repeated X-ray burst.

8. The Rapid Burster – None of the Above

One particular source, the *Rapid Burster*, displays behavior that falls in the "none of the above" category. This system, known to intimates as MXB 1730–335 (for MIT X-ray Burst), was discovered about 20 years ago. When active, it bursts about four thousand times a day. The Rapid Burster is located in a globular cluster. It also occasionally has the more prominent bursts associated with the thermonuclear ignition of helium. Like the other thermonuclear burst sources, the Rapid Burster shows no sign of X-ray pulses that would indicate the rotation of the underlying neutron star. The presumption is that the magnetic field of this neutron star is relatively weak so matter falls more uniformly on the surface and is not focused at the magnetic poles. The repetitive bursts that define the Rapid Burster are thought to be neither a thermonuclear burst on the surface of the neutron star nor the type of accretion disk heating instability similar to that of dwarf novae. The observations suggest that the matter rains down on the neutron star in blobs like a rapidly dripping faucet, rather than in a steady gush. There is no well-established theory for this behavior, but the suspicion is that it involves an instability of the matter on the inner edge of the accretion disk that may be due to a condition where the pressure of radiation becomes excessively large, larger than the pressure of the hot gas in the disk. For 20 years, the Rapid Burster was alone, but now it has some company.

In 1990, NASA launched another of its great observatories to complement the *Hubble Space Telescope*. This was the *Compton Gamma-Ray Observatory*. We will talk about it more in Chapter 11. In December 1995, this satellite discovered a system known as the *bursting pulsar*, or, more technically according to its discovery instrument and coordinates, GRO J1744–28. Follow-up work on it was done by another NASA satellite, the *Rossi X-ray Timing Explorer*. This relatively modest satellite was named after Bruno Rossi, an MIT pioneer of X-ray astronomy, and was designed to follow X-ray behavior with very accurate timing. Observations with *RXTE* of the bursting pulsar showed an incredible array of behavior that indicate that this system may be an important link between systems like the Rapid Burster and the other X-ray burst sources.

As its name implies, the bursting pulsar is an X-ray pulsar. From the frequency of the pulses, one can deduce that the neutron star rotates about twice a second. Its orbital motion has also been detected. The neutron star is in a 12-day orbit around a small red giant that has lost almost all of its hydrogen envelope and now has a mass of about one-quarter the mass of the Sun. From January through May of 1996, the bursting pulsar showed large bursts lasting about 10 seconds apiece every 2 hours or so. These bursts displayed character-istics of the staccato bursts of the Rapid Burster rather than the helium ignition flares of the X-ray bursters. The presumption is that the bursting pulsar has a stronger magnetic field than the Rapid Burster and hence can both generate "lighthouse" pulses of X-rays from the magnetic poles and can suppress nuclear flares by the fo-cused, hot accretion at the magnetic poles. The fact that it still man-ages to show the instability of the inner disk means that the magnetic field is not so strong that it cuts out the inner part of the disk where that instability happens. The bursting pulsar is thus an interesting intermediate case that promises to teach us more about the conditions under which neutron stars evolve in binary systems. After May of 1996, the system got so dim that *RXTE* could no longer detect it, so for now, the bursting pulsar is keeping any further secrets it may have to reveal.

9. Millisecond Pulsars

In the last decade, a new variety of radio pulsars have been found that have generated great excitement because they link so many as-pects of the formation and evolution of neutron stars. Theory predicts that neutron stars cannot rotate faster than about one thousand times per second without flinging themselves apart with the excessive cen-trifugal force. That limiting rotation rate corresponds to a rotational period of 0.001 second, or 1 millisecond. Thus one expects that the fastest pulses that could be discovered from a pulsar would be about 1 millisecond, and that a 1 millisecond pulsar would be on the verge of tearing itself apart. Realistically, one would expect that pulsars would rotate a little slower than this fastest possible limit and, hence, to have pulses of a few milliseconds. By this standard, the pulse period of the Crab nebula pulsar is dawdling along at a mere 33 milliseconds.

Special search techniques were developed to search for pulsars near this period limit, and they have been successful. Over two dozen *millisecond pulsars* have been found. In contrast to their longer pe-riod kin, about half of the millisecond pulsars are in binaries. The most rapidly rotating has a remarkably well-defined period, 0.00155780644885 seconds, or about 1.6 milliseconds. This neutron star is whipping around 642 times per second.

The next step is to account for the origin of the millisecond pulsars.

Pulsars must be magnetic neutron stars. The Crab pulsar rotates 30 times per second; normal pulsars, about once per second. This is because the Crab pulsar is only 1,000 years old. When it is several million years old, the Crab pulsar will have slowed down, and it will presumably also have a period of about 1 second. This suggests that millisecond pulsars might be very young, newly born neutron stars. More thought, and appropriate observation, shows just the opposite is the case. With a normal-strength magnetic field, a pulsar with a period of 1 millisecond would be losing energy so fast that it could not maintain its rapid rotation. By this argument, the millisecond pulsars should be slowing down very rapidly, but they are observed to be slowing scarcely at all. The millisecond pulsars must therefore have a smaller magnetic field than normal so that they lose little rotational energy into radiation. This in turn suggests that they are old so that there has been time for their magnetic fields to decay away or otherwise disappear. If they are old, however, why have they not lost more of their rotational energy when they were younger with a more robust magnetic field?

The proposed resolution to this query is that the neutron stars were born in binary systems and that transfer of mass and associated angular momentum from a companion kept the neutron star spinning fast even as the field decayed. Thus all millisecond pulsars should be in binary systems, but a significant fraction of them are not. This is another dilemma. If there were a binary companion, where did it go?

One possible answer to this further dilemma was suggested by the discovery of a particular millisecond pulsar in a binary system. This pulsar orbited a companion, a more or less normal star. It appeared as if the pulsar were killing the normal star because the star was losing mass at a high rate. The rapidly rotating neutron star produces a great flux of high-energy radiation, X-rays and gamma rays. It was first thought that this intense radiation was literally blasting away the companion star. Some astronomers termed this system the Black Widow star because the neutron star was perceived to be killing its mate. Subsequent observations showed that the star was probably losing mass on its own. In any case, the implication is that the companion will soon be gone, leaving a millisecond pulsar to spin alone in space. Roughly half of the millisecond pulsars are in binary systems with a companion star to transfer mass and keep them spun up. Presumably the other half of the observed millisecond pulsars have already dispensed with their companions in one way or another.

Another interesting millisecond pulsar revealed that it had objects of planetary mass orbiting it. These objects were discovered only by the exquisite timing that is possible with these pulsars. Tiny rhythmic oscillations in the pulse period revealed that the pulsar was being slightly tugged around in space by several small objects of mass about that of Jupiter. Whether these are true planets, left over from some ill-fated solar system that orbited the star before it exploded, or whether the "planets" are themselves left over lumps of blasted star-stuff is not clear.

To put the millisecond pulsars in perspective, we need to take a step back in the evolutionary story. What sort of system gave rise to the original system of a neutron star orbiting an ordinary star? The explosion of a supernova in a binary system ejects a great deal of mass and hence decreases the gravity that holds a binary system together. That is why we think most ordinary pulsars are alone in space. They have not murdered their companions, but they may have unbound and ejected them from orbit. To prevent this, we need a fairly gentle way to make a neutron star. After the neutron star is born, it must have a weak magnetic field or lose an originally strong magnetic field and then be spun up by accretion to become a millisecond pulsar.

If this is the evolution of the neutron stars that become millisecond pulsars, then such systems should pass through a phase in which the companion adds mass to the neutron star to spin it up. The result should be the production of X-rays. The natural conclusion is that the systems we see now as X-ray sources with neutron stars orbiting low-mass companion stars will evolve to become the millisecond pulsars. The problem is that if you work out the rate at which X-ray systems with neutron stars and low-mass companions are born and the rate at which millisecond pulsars are born, they disagree substantially. There do not seem to be enough low-mass X-ray systems to account for the number of millisecond pulsars. Either there is another way to make millisecond pulsars, or there is something we do not understand about the evolution of the stellar systems in the X-ray phase. If that phase lasted a shorter time than we think, there would have to be a higher production rate to account for the number we see at this epoch in galactic history. That would help close the gap.

Another mechanism that might avoid the phase of being an ordinary X-ray source during the spin-up phase has been suggested to produce millisecond pulsars. That mechanism involves the accretion of matter onto the O/Ne/Mg core of a star of original mass of about

10 solar masses. When such a core reaches its maximum mass, it will undergo electron capture and collapse to form a neutron star, but essentially all the core will collapse to make the neutron star, and very little is expected to be ejected (Chapter 6). This gives the maximum probability of maintaining a companion in binary orbit. This general process is called *accretion induced collapse*, to distinguish it from core collapse brought on by the normal process of core collapse of a single evolving star as fuel is burned to heavier elements. This process is plausible in general, but it does not necessarily predict that the resulting neutron star will be rapidly spinning with a low magnetic field, the conditions required to be a millisecond pulsar.

The low magnetic fields required to explain the millisecond pulsars have raised a different conundrum. All radio pulsars are observed to fall on the short period side of a limiting value of the period that depends on the strength of the magnetic field. The implication is that as pulsars age and rotate slower and slower, their magnetic fields decay away so that for very old slow pulsars the combination of rotation and magnetic field is no longer able to generate the thunderstorms at the magnetic poles that are required to make a radio pulsar. In a plot of magnetic field versus spin period, this limiting period is known as the "death line." Taking a somewhat more pragmatic approach, Mal Ruderman of Columbia University argues that the cutoff may be different for different magnetic field configurations and hence the boundary may be a "death valley." In any case, the notion persisted for two decades that the magnetic field of pulsars decays away with a time scale of perhaps 100 million years. Continued consideration of the numbers of pulsars with different field strengths and spin periods and the existence of the millisecond pulsars with very low magnetic fields has inspired reconsideration of this issue. There are suggestions that the field may not decay or that it is the accretion process itself that kills the field in the case of the millisecond pulsars. The origin and evolution of neutron star magnetic fields is still a subject of active investigation.

10. Soft Gamma-Ray Repeaters – Reach Out and Touch Someone

Although the Sun occasionally belches a flare of particles that reach the Earth and affect radio communications, we are used to the stars being quietly remote in their isolated magnificence against the backdrop of dark space. Imagine our surprise, therefore, when one of them reached out and touched us in August of 1998! As the Earth sails around the Sun and follows the Sun around the Galaxy for billions of years, it is not isolated from the violent Universe around us.

A class of bursting events called *soft gamma-ray repeaters* has been defined over the last 20 years. At first these events were confused and intermingled with the events known as *gamma-ray bursts*, the story of which we will learn in Chapter 11. The difference between "hard," high-energy X-rays and "soft," low-energy gamma rays is a matter of operational definition, and the dividing line is somewhat arbitrary. As the names imply, however, soft gamma-ray repeaters and gamma-ray bursts radiate most of their energy in the gamma-ray range. The soft gamma-ray repeaters emit somewhat less energetic photons than the gamma-ray bursts, a difference an expert can love. As we shall see in Chapter 11, no gamma-ray burst has ever been known to repeat. As data accumulated, however, it became clear that the sources that gave out the softer gamma-rays could and did repeat their outbursts, if at irregular intervals. The question was, what were they? Gamma rays of any sort require high energies, and that suggests high gravity, so one might think about white dwarfs, neutron stars, or black holes. Round up the usual suspects! An important clue was that all the soft gamma-ray repeaters turned out to be in supernova remnants.

The currently most widely accepted theory for the soft gamma-ray repeaters was developed by Rob Duncan at the University of Texas and Chris Thompson now at the University of North Carolina. They were originally seeking an explanation for gamma-ray bursts, not soft gamma-ray repeaters. Their investigations led them to consider neutron stars with very strong magnetic fields. They developed a theory that, under certain circumstances involving, among other things, very rapid rotation, neutron stars could develop immensely strong magnetic fields. Whereas millisecond pulsars have magnetic fields about ten thousand times less strong than "normal" pulsars, Duncan and Thompson argued for magnetic fields thousands of times stronger than "normal." The force of such magnetic fields could rival the gravity of the neutron star – strong indeed. Duncan and Thompson needed a name to distinguish their intellectual baby from the "normal" pulsars and millisecond pulsars, so they coined the phrase *magnetar* for a neutron star where the magnetic field rivaled gravity and pressure.

As they investigated the properties of magnetars, Duncan and

Thompson realized that they should have a special activity. When they are first born, the magnetars would assume an equilibrium balancing the magnetic fields, pressure, gravity, and the centrifugal force of their rapid rotation. The latter would cause the neutron star to bulge along the equator, and that bulge would tend to be frozen into place in the outer rocky crust of the neutron star. As the neutron star lost energy and slowed, the bulge would be too big for the slower rotation, and it would eventually crack and settle. This picture is very similar to the original explanation for glitches in pulsar rotation rates, which has now been supplanted, as mentioned in Section 4. In the context of the magnetar theory, however, Duncan and Thompson realized that such a crust cracking would send powerful waves into the magnetic field that looped above the neutron star surface. The magnetic field would have to readjust to the new structure of the neutron star, and the magnetic field would convert some energy into hot plasma. That hot plasma would radiate the gamma-ray energy for the time scales observed in soft gamma-ray repeaters. Duncan and Thompson proposed that soft gamma-ray repeaters were, in fact, magnetars, a variety of super magnetized neutron star not previously recognized. They also recognized that after the first, major, crust-cracking star quake, there could be more localized shifts in the crust as it adjusted to the rearranged magnetic field. This would give a smaller, dimmer source of soft gamma rays, but if the spot were carried around by the rotation of the neutron star, then one might see a "lighthouse" effect so that the gamma rays would be seen to "pulse" at the rotation rate of the neutron star.

This suggestion that soft gamma-ray repeaters were magnetars attracted some positive, some negative, and some bewildered reactions. To make progress, observational confirmation was needed, and that came in 1998 in a rapid succession of events. Careful observations with the *Rossi X-ray Timing Explorer* revealed the rotation rate and rate of slowing down of one of the soft gamma-ray repeaters. The observations were consistent with a neutron star with a magnetic field one thousand times stronger than "normal."

In August of 1998, Nature made sure we understood this lesson. One of the soft gamma-ray repeaters went off with a burst that was so strong that it affected the Earth! The gamma rays from this soft gamma-ray repeater affected the ionization of the upper atmosphere and interfered with radio communications worldwide. A wonderful contribution to the Op/Ed page of the *New York Times* described the awe-inspiring, incredibly intense, and widespread aurora witnessed by a bunch of guys on a fishing expedition above the Arctic Circle. This was the first known event when a star beyond the Solar System physically affected the Earth. There was no harm done, but this cannot have been the first time such a thing happened, and it will not be the last.

The event just described also brought evidence for a pulsar with a superstrong magnetic field. It had the immensely strong burst that tickled the Earth's ionosphere, but then a series of pulses just as Duncan and Thompson had predicted for the subsequent hot spots that should occur as the crust shifted in spots. In hindsight, just this behavior had been seen in the first soft gamma-ray repeater observed in 1979 in the Large Magellanic Cloud. At the time, that outburst was strange and controversial. That misery is now comforted by the company of the nearly twin outburst of the nearby source that produced the August 1998 burst. One must be careful and continue to seek evidence, but the magnetar theory is clearly the leading contender to account for the soft gamma-ray repeaters.

There is a handful of other objects that also seem to fit nicely into this scheme. These have been known as the *anomalous X-ray pulsars*. Like the soft gamma-ray repeaters, the anomalous X-ray pulsars are all found in supernova remnants. They show no evidence for binary companions. They all have rather long periods that fall in a restricted range of 6–11 seconds, very similar to the soft gamma-ray repeaters. They all seem to be spinning down, the spin periods getting longer and longer as if the spinning source were simply losing energy. From the spin period and rate of decrease of spin, an indirect estimate can be made of the strength of the magnetic field and the result is a value comparable to magnetars: 100 to 1,000 times stronger than normal radio pulsars.

A scheme that makes sense is that one neutron star in ten is born with an especially high rotation that allows the newly born neutron star to generate the high magnetic field. For the next 1,000 years, that magnetar undergoes crust cracks and rearrangement and is active as a soft gamma-ray repeater. After that time, the neutron star rotates sufficiently slowly that it cannot generate strong gamma-ray outbursts, but for the next 40,000 years it can radiate enough to be seen as an anomalous X-ray pulsar. After that time, it will be cooler and slower and will be a "dead" magnetar. The nature of the supernova that gives rise to magnetars and the nature of dead magnetars are not clear. How often do we end topics on that note? Such a big Universe, so little time. . . .

11. Geminga

Yet another chapter in the neutron star story is told in the saga of the source known as *Geminga*. This source was first detected by one of the early satellites with gamma-ray instruments in 1973. Two decades were required to figure out what it was. The name was given to it by an Italian X-ray astronomer, Giovanni Bignami. The name is nominally related to the fact that it is a gamma-ray source in the direction of the constellation of Gemini. More amusingly, it is an Italian double entendre related to the fact that the source could not be detected in the radio, one of the on-going mysteries. In the dialect of Milan spoken by Bignami, *ghe'è minga* means it's not there.

Vision in gamma rays is blurry and there were lots of spots of light in the direction of Geminga. A long time was required to pin down the source. In the optical, stars, asteroids, and plate defects had to be ruled out. The *Einstein* satellite revealed an X-ray source that helped to narrow down the optical search. One thing became clear. Whatever the object was, it was damn dim in the optical. Suspicion that Geminga was a neutron star grew. In the late 1980s, a dim optical source was isolated. It turned out to be the real thing.

A major breakthrough came finally in the 1990s with observations from the *Compton Gamma-Ray Observatory* and the *German Röntgen Satellite* or *ROSAT*, named for the discoverer of X-rays. Observations with these instruments showed that Geminga revealed both gamma-ray and X-ray pulses due to rotation with a period of 0.237 seconds. Geminga was a neutron star. Like the Crab nebula pulsar it emitted gamma rays, but unlike the Crab pulsar and so many others, it did not emit radio radiation. Various arguments suggested that it was very close to the Earth. That meant that, even though the gamma rays were detected, they were intrinsically feeble. That was why similar sources were not common. They would just be too hard to detect at greater distances. The small distance also explained why Geminga could be seen in the optical at all. Neutron stars have such a small radiating surface that one would have to be very close to be observed.

The close distance had another significant implication. There was a chance to detect the *proper motion* of the source, the motion across the sky due to its motion through space, and even the *parallax*, the apparent motion due to the Earth's orbit around the Sun. The former gives a hint of where Geminga arose; the latter, how far away it is. The parallax was measured in 1994 with the *Hubble Space Telescope*, and Geminga is only 160 parsecs, about 500 light years away – right in our back yard! The proper motion was extrapolated backward, and Geminga's origin was traced to near a star in the Orion nebula. There is an expanding cloud of gas around a star there that might be the supernova that created Geminga. The time for Geminga to get from Orion to where it is now is about 350,000 years, which is consistent with the age measured from the rate of slowing of the spin and with the estimated age of the supernova remnant. There are other possible interpretations, but the strong implication is that Geminga arose in a supernova explosion rather nearby about 350,000 years ago. Early hominids were leaving the veldt then and beginning to explore the planet – not so long ago.

The interpretation of Geminga is that it is a neutron star with a rather normal magnetic field. In 350,000 years, it has spun down so that it can barely generate gamma rays by particle creation and acceleration near the magnetic poles. Its surface is still hot and glows in the optical, if dimly. The most likely reason why radio is not observed is that the radio is created, but that it is radiated away from the Earth by an accident of orientation. Overall, Geminga is very special because of its nearness to Earth, but it may represent a normal phase in the aging and evolution of normal neutron stars. In looking to the past of Geminga, we may also be looking to the future when Betelgeuse erupts at about the same distance, the story foretold in the sidebar in Chapter 6.

9

Black Holes in Theory

Into the Abyss

1. Why Black Holes?

Black holes have become a cultural icon. Although few people understand the physical and mathematical innards of the black holes that Einstein's equations reveal, nearly everyone understands the symbolism of black holes as yawning maws that swallow everything and let nothing out. Can there be any compelling reason to understand more deeply a trivialized cultural metaphor? The answer, for anyone interested in the nature of the world around us, is an emphatic yes! Black holes represent far more than a simple metaphor for loss and despair. Although black holes may form from stars, they are not stars. They are objects of pure space and time that have transcended their stellar birthright. The first glimmers of the possibility of black holes arose in the eighteenth century. Two hundred years later, they are still on the forefront of science. In the domain of astronomy, there is virtual certainty that astronomers have detected black holes, that they are a reality in our Universe. In the domain of physics, black holes are on the vanguard of intellectual thought. They play a unique and central role in the quest to develop a "theory of everything," a deeper comprehension of the essence of space and time, an understanding of the origin and fate of our Universe.

There is a certain inevitability to black holes in a gravitating Universe. Einstein's theory says that for sufficiently compressed matter, gravity will overwhelm all other forces. The reason lies in the fundamental equation, $E = mc^2$. Because mass and energy are interchangeable, one of the implications of this equation is that energy has weight. The very energy that is expended to provide the pressure to support a star against gravity increases the pull of the gravitational field. The more you resist gravity, the more you add to its strength. The result is that if an object is compressed enough, gravity becomes overwhelming. Any force that tries to resist just makes the pull all the greater. When gravity exceeds all other forces, the object will collapse to form a black hole.

The first people to contemplate the notion that gravity could become an overwhelming influence were John Mitchell, a British physicist, and the Marquis de LaPlace, a French mathematician. Mitchell in 1783 and LaPlace in 1796 based their arguments on Newton's theory of gravity. They used the concept of an *escape velocity*. The notion is that to escape from the surface of a gravitating object, a sufficiently large velocity must be imposed to overcome the pull of gravity and "escape" into space. If the velocity is too small, the launch will fail. If it is just right, a launched vehicle will just coast to a halt as it gets far away from the gravitating object. With more velocity, a launched vehicle will still have a head of steam as it breaks free of gravity and it will continue to speed away. That is the whole idea behind tying two big, solid-fueled boosters and an external liquid fuel tank to the space shuttle when it goes up from Cape Canaveral. The shuttle must achieve escape velocity, or near it, to get into orbit, and that means lifting off the launch pad really fast!

Mitchell and LaPlace used this idea of an escape velocity to argue that an object could be so compact that the escape velocity from the surface would exceed the speed of light. By some coincidence, an algebraic formulation of this escape velocity condition in the context of Newton's theory of gravity gives the correct result for the "size" of a black hole using the correct theory of gravity, general relativity. Mitchell did not, apparently, coin a zippy shorthand name for his intellectual creation. LaPlace called his hypothetical compressed entities *corps obscurs*, or hidden bodies. (The modern French equivalent is *astres occlus*, or closed stars. The literal translation, *trous noirs*, has also gained acceptance after some initial resistance because of its suggestion of double entendre.)

With some hindsight, we can see that Newton's theory of gravity

was flawed. This theory predicted that, if two masses got infinitesi-
mally close together, the force would go to infinity. A general lesson
of physics is that, when infinities arise, there is a problem with the
mathematical formulation that reflects some omission in the physics.
Another problem with Newton's law of gravity is that, although it
prescribed how the strength of gravity scaled with the mass of a
gravitating object (to the first power) and the distance between objects
(inversely as the square of the distance), it did not say how gravity
varied in time. Consider two orbiting stars. A literal use of Newton's
law of gravity says that, as one star moves, the other instantaneously
responds to the fact that the motion has occurred. Thus according to
Newton's law of gravity, the effect of gravity propagates infinitely
fast. This second troublesome infinity violates the idea that nothing
can move faster than the speed of light. Finally, and perhaps most
compelling from a strictly practical point of view, Newton's gravity
did not work.

Newton's law of gravity is spectacularly successful in most normal
circumstances, when distances are large and speeds are slow. Astronomers still use it to great effect to predict the orbits of most stars.
Rocket scientists use it to plot the paths of spacecraft even as they do
complex orbits that carry them around planets, getting a boost from
the interaction. The *Galileo* spacecraft went through a remarkable
series of bank shots around the inner planets, picking up speed in the
various encounters with Venus and Earth, before being flung to Jupiter. The recently launched *Cassini* spacecraft completed the first stage
of its voyage to Saturn by first looping inward to circle Venus. *Cassini*
received a kick from the orbital motion of Venus that gave it the momentum to sail out to Saturn. The success of gravitational multiple-
bank shots shows that Newton's gravity works very well in this regime.

For very fine measurements, however, Newton gives the wrong
answer! The predictions of Newton do not agree with observation,
with the way nature works. Classic examples are the rate of rotation
of the perihelion of Mercury and the deflection of light by the Sun. In
contrast, Einstein's theory of gravity has passed every test of observation. A modern example is the use of global positioning systems
(GPS) in boating, camping, and driving as well as military and industrial uses. This system works by timing the signals from an array of
orbiting satellites. It is based on the mathematics of the curved space
and warped time of Einstein, not the simple law of gravitation of

Newton. If the silicon chips in the GPS detectors knew only about Newton, boaters in the fog and soldiers in the field would get lost!

As we shall see, giving up Newton for Einstein does not represent merely swapping one set of mathematics for another. Rather, Einstein brought with him a revolution in the fundamental concept underlying gravity. Newton crafted his mathematics in the language of a force of gravity as the underlying concept. Physicists and astronomers still use the notion of a gravitational force in casual terms, even though it has become outmoded in a fundamental way. Einstein's view was radically different. For Einstein, there is no force of gravity. Instead, Einstein's theory represents gravity as a manifestation of curved space. A gravitating object curves the space around it. A second object then responds by moving as straight as it can in that curved space. The curved space results in deflections of motion that are manifested as gravity even though the object is in free fall, sensing no force whatsoever. Much of this chapter will be devoted to exploring this conception of gravity.

The progress of our understanding of gravity is not over, however. We have come to understand that, even though it has passed every experimental test, Einstein's theory has flaws. It has its own nasty infinities that represent some omission in the physics. Ironically the hints of a new, better theory are again cast in the language of force, but not the force of Newton. In notions being developed today, the force is quantum in nature and may play on a field of ten or eleven dimensions, not the three of space and one of time that sufficed for both Newton and Einstein. We will begin with an exploration of black holes as portrayed in Einstein's theory and see how deeper issues arise. Some of those issues will be explored in Chapter 12.

2. The Event Horizon

As described by general relativity, a black hole is a region of space-time bordered by a one-way membrane called an *event horizon*, as shown schematically in Figure 9.1. Matter or light can pass inward through the event horizon, but nothing that travels at or less than the speed of light, even light itself, can get back out. The term "event horizon" comes from the notion that if an "event," like a firecracker exploding, occurs just outside the event horizon, the light can reach an observer, and the fact that the event occurred can be registered. If

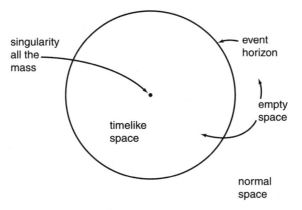

Figure 9.1. The simplest, nonrotating black hole has two basic elements – the event horizon, interior to which nothing can escape, and the singularity, where everything, including space and time, are crushed out of existence. Within the event horizon, space takes on a timelike aspect (Section 8.1).

the firecracker goes off just inside the event horizon, however, no information that the event occurred can reach the observer. The event takes place beyond a horizon so that it cannot be seen. Once inside the event horizon, escape is impossible without traveling faster than the velocity of light. The location of the event horizon is thus intimately related to the fact that the speed of light is a speed limit for all normal stuff. The simple argument of Mitchell and LaPlace concerning the formation of a *corps obscurs* relates to the size of the event horizon. The size of the event horizon scales with the mass of the black hole. For a black hole with ten times the mass of the Sun, it would have a radius of 30 kilometers, about 50 miles in diameter. The nature of the event horizon in the context of curved space and time will be explored in more depth in Section 5.

3. Singularity

When Newton was pondering the means by which apples bonked him on the head and, more particularly, how the Earth kept the Moon trapped in orbit, he intuited an important aspect of gravity. He realized that the gravity of the Earth must act from the center of the Earth, not, for instance, from its surface. This was not a trivial conclusion, and he needed to prove that it was true. Newton knocked off his gravity studies for a while and invented the mathematics of calculus in order to prove his conjecture. With his new mathematical tools, Newton was able to prove that, although the mass of the Earth is distributed throughout its volume, each little piece of the Earth acts in concert as if it were in the center. The result is that for any object beyond the Earth's surface, the gravitational attraction of the Earth will act as if all the mass of the Earth were concentrated at a point in the center. This is true for any spherical gravitating body. The gravitational attraction depends only on the distance from the center of the body, not on the radius or volume of that body. Armed with this mathematically proven conclusion, Newton went on to formulate his theory of gravity with a mathematical expression that said that the force of gravity between two spherical objects depended only on the masses of the two objects and on the inverse square of the distance between their centers.

As an example to make this property concrete, imagine that the Sun were suddenly compacted to become a neutron star of the same mass. It would get cold and dark on the Earth, but the Earth would continue in exactly the same orbit because the gravitational pull it feels from the Sun depends only on the mass of the Sun, not on how big it is. Another implication is that we are in no danger of falling into a black hole. All the black holes we know or suspect are far away. The gravity would be frightful if we were to get near their centers, but at a large distance from their centers, the gravity gets weak as it does at a large distance from any object and vanishingly small if the distance is very large. In this context, there is one interesting difference between normal stars of any kind – suns, white dwarfs, or neutron stars – and black holes. The former act as if all their mass were concentrated at a point in the center. For black holes, this is literally true.

Inside the event horizon, all mass that falls into a black hole is trapped. Even though there is no material surface at the event horizon, the matter within the black hole still signifies its presence by exerting a gravitational pull. The gravitational acceleration exists outside the event horizon and causes the formation of the event horizon. Although the black hole still exerts a gravitational pull, the matter itself is crushed out of all recognizable existence. General relativity predicts that the matter compacts into a region of zero volume and infinite density at the center of the black hole. Even more profound, space and time cease to exist at this point. Such a region is called a *singularity* and is illustrated schematically as a point in Figure 9.1. For a black hole, all the mass that creates the gravity is literally at this point in the center, at the singularity.

The infinities associated with the singularity are clues that Einstein's theory is not a complete theory of gravity, despite its great success. We know in principle what is lacking. Einstein's theory does not contain any aspects of the quantum theory. The uncertainty principle of the quantum theory tells us that it is not possible to specify the position of anything exactly, including the position of an infinitely small singularity. The notion of a singularity as it arises in Einstein's theory is thus an intrinsic violation of the quantum theory. With a theory of gravity that properly incorporated quantum effects, which general relativity does not, the singularity would probably be altered to be a region of exceedingly small volume and immense, but not infinite, density. It is the nature of that exceedingly small volume, the singularity that forms inside a black hole, the singularity from which our Universe was born, that is the heart of the quest for a new, deeper understanding of physics.

4. Being a Treatise on the General Nature of Death Within a Black Hole

The manner in which a black hole crushes matter out of existence, save for its gravitational field, is rather graphic. Consider something falling into a black hole, say a human body – feet first. In this case, at every instant the feet are going to be closer to the center of the black hole than is the head. Gravity is thus going to be stronger at the feet and will pull the feet away from the head. The natural forces on an extended body tend to stretch it along the direction toward the center of the gravitation. At the same time, all parts of the body are trying to fall toward the center. The left shoulder is trying to fall toward the center. The right shoulder is trying to fall toward the center. As the body gets closer to the center, the distance between separate paths directed at the center gets ever smaller. The shoulders get shoved together, and whatever is in between must suffer the consequences. A body falling into a black hole will be stretched feet from head and crushed side to side. This is known jocularly as the "noodle effect." Anything falling into a black hole will be noodleized, as shown in Figure 9.2.

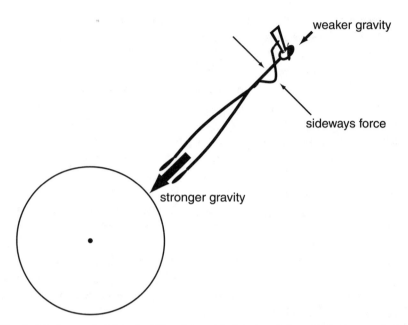

Figure 9.2. Any material body falling into a black hole will have its bottom pulled from its top and its sides crushed together in a tidal "noodleizing" effect.

The technical name for this simultaneous radial stretching and lateral crushing is the *tidal force*. It is precisely the same effect as causes the tides on the Earth. Here the Moon pulls on the Earth and its oceans, pulling them toward the Moon and pushing them in sideways to form the tidal bulges in the oceans, the faintest form of noodle. As a body falls into a black hole, the tidal forces increase drastically. First the body stretches into a noodle and breaks apart. Then the individual cells stretch into noodles and are destroyed. Next gravity overcomes the electrical forces that bind matter into molecules and atoms. Atoms will be wrenched out of molecules and electrons pulled from atoms. As infall proceeds, the rising tidal forces will overcome the nuclear force, stretching out the atomic nuclei and breaking them apart into individual protons and neutrons. In their turn, the protons and neutrons will break up into quarks, and the quarks into whatever comprises them. These building blocks will in turn be subject to supernoodlization until the singularity is reached and matter as we know it ceases to exist. Another way of characterizing the singularity in Einstein's theory is that the tidal forces become infinite. Physicists are gaining the first hints of what conditions may be in the singularity that will prevent that infinity. A discussion of this topic is postponed to Chapter 12.

5. Black Holes in Space and Time

5.1. Curved Space and Black Holes

Black holes are in the most fundamental way a beast of curved space. Visualizing this curvature that occupies all of three dimensions is very difficult for creatures such as us who are limited to a three-dimensional perspective. Even the experts have difficulty picturing the immense complexity of curved space. They have invented tricks to help with the perception. We will describe these tricks because they help, but even they represent only a shadow, and a fairly complicated one, of the truth.

The notion of curved space raises a general question. How do we characterize it? A line inscribed in a wavy two-dimensional space may be straight in some sense from our three-dimensional perspective, but not truly straight at all. Likewise, a properly "straight" line in a curved two-dimensional space may look strangely curved from an-other perspective. The ability to define and construct straight lines in curved space is fundamental to understanding how curved space works.

What do we mean by a straight line in curved space? There is a rigorous way to decide which lines are straight in a given space, a way that is intuitively reasonable as well. To obtain a straight line in a curved space, start with a small portion of the space where it is, for all practical purposes, flat. Think of any measurement you would normally make on the surface of the Earth, ignoring the fact that the Earth is really a closed spherical surface. In this small, nearly flat portion, use two short straight sticks. Lie one stick down. Now extend the second stick so that it partially overlaps the first, so that you know it is pointed in the same direction as the first, but so that it also extends out a way. Now hold down the second stick and slide the first along, keeping it parallel to the second stick until it extends out a way. Continue in this manner, extending each stick in turn a little way in such a manner that you are always assured that each extension goes in precisely the same direction as the last. As you proceed, draw a line using each stick in turn as a straight edge. Never look off at a distance to orient yourself. This technique depends on the fact that you are looking only at the local little patch of very nearly flat space in which you find yourself at any given instant. This method of drawing a straight line is called *parallel propagation* because each step consists of extending one of the sticks parallel to the other. One can prove mathematically that the line you draw as a result of this tedious operation is the shortest distance between any two points along it. What more could you want from a truly straight line? The operation of parallel propagation is what you approximate every time you sketch a freehand straight line. You do not make two marks on a paper and then try to make the distance between them as short as possible. Rather you start your pencil off in some direction and then, trying to keep your hand steady, continue the line parallel to itself. That is what makes parallel propagation so intuitive. It is what you really do to sketch a straight line.

In a flat space, parallel propagation will give the ordinary straight lines known and loved by tenth-grade geometry teachers. Parallel lines constructed in this fashion will never cross. Triangles made of three such lines will have 180 degrees as the sum of their interior angles. This is the geometry of Euclid, the geometry of flat space. In an arbitrarily curved space, watch out! Viewed from above, lines drawn as straight as possible by the method of parallel propagation will appear wackily curved if the surface is curved; but parallel propagated lines are as straight as possible and will be the shortest distance between two points even if the space is curved.

A particular trick the mathematicians have developed for picturing curved space is to project a three-dimensional curved space onto two dimensions in a special way, like casting a shadow. One dimension is suppressed, and the resulting two-dimensional figure is displayed as a two-dimensional surface in three-dimensional space. It becomes something we can look over, around, and under from our three-dimensional perspective and get a feel for the real thing. The technical name for the image that results from projecting the two-dimensional representation into ordinary flat, three-dimensional space is called an *embedding diagram*, because the two-dimensional "shadow" is embedded in the three-dimensional space.

To perform this trick for a black hole, one of the dimensions of rotation is suppressed. The resulting figure looks like a cone, or as if you were to poke your finger into a rubber sheet, as shown in Figure 9.3. The distant, still flat, parts of the sheet are the simple two-dimensional projection of flat, uncurved, three-dimensional space.

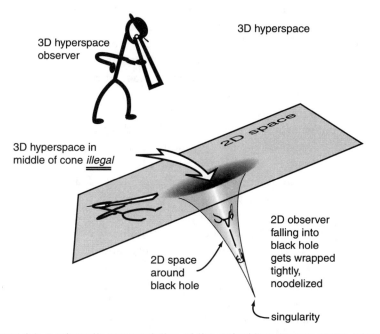

Far from 3D hyperspace observer

3D hyperspace observer

3D hyperspace

3D hyperspace in middle of cone *illegal*

2D space

2D space around black hole

2D observer falling into black hole gets wrapped tightly, noodelized

singularity

Figure 9.3. A schematic representation of the embedding diagram of the curved, two-dimensional space around a black hole. Far from the black hole the space is flat. Near the black hole, the space appears to be a "cone" to a three-dimensional hyperspace observer. A two-dimensional scientist falling into the black hole would be stretched toward the singularity, wrapped in the conical space, and crushed in the singularity. Note that in this view, the space corresponding to the two-dimensional black hole is on the cone. The region "within" the cone as perceived by the hyperspace observer is part of the higher, three-dimensional space that is imperceivable and inaccessible to a two-dimensional inhabitant of the two-dimensional space.

The cone made with your finger is a technically proper representation of the curved space around a black hole (at least in qualitative shape, the mathematics of Einstein's theory tells the precise shape of the cone).

Full appreciation of the manner in which this cone represents the curved space of a black hole takes some time and quiet contemplation. One feature of the cone is immediately apparent and quite important. Consider the construction of a circle on the surface around the cone. This operation must be done in the confines of the two-dimensional surface. To go off this surface into three dimensions is cheating because that would be like going from the real three dimensions of a black hole into an unphysical honest-to-gosh fourth spatial dimension. To draw a circle, start at the center of the "black hole," at the bottom of the depression of the cone. Draw a line out along the curved surface directly away from the center. This line is a radius line, despite the fact that, from our three-dimensional view of the operation, it follows the funny curved surface of the cone. Now stop at some point along the surface of the cone and draw a circle, a line connecting all those points that are equally distant from the center.

Now imagine that you measure the length of the radius line and the circumference of the corresponding circle. Do you see that the radius in this curved surface must always be longer than normal? The ratio of the circumference to 2π times the radius is always less than one. The process of constructing the cone preserves this aspect of the original curved space, and the resulting embedding diagram lets it be seen graphically. In this curved space, the distance inward as represented by the radius is somehow stretched and lengthened. If you were to go off to a flat portion of the rubber sheet and do the same operation, start at a point, go out a certain distance along a radius, make a circle, you would get the standard result – the circumference is 2π times the radius. That is the test for flat space.

Let us apply the technique of parallel propagation to the curved space around a black hole as portrayed by the projected two-dimensional cone, as illustrated in Figure 9.4. Figure 9.4 shows two scientists drawing lines by parallel propagation in the two-dimensional space they occupy. Both start at some distance out in the "flat" portion. One draws a parallel-propagated line that passes far from the black hole. This line looks straight to an imaginary three-dimensional hyperspace observer, the perspective we take whenever we look down from our three-dimensional hyperspace onto a two-dimensional embedding diagram. The other scientist draws a parallel-propagated straight line that skirts the deepest portion of the cone (we do not want anyone crushed by the infinite tidal forces!). As this line nears the lowest portion of the cone, think what happens. A small portion of the space surrounding this point is oriented differently than a small portion of the space out in the flat, away from the cone. The line drawn in this location is going around the axis of the cone, responding to the "aroundness" of the surface, despite the fact that it is going as straight as it can in the curved space of the cone. From this part of space, the line must head off in a direction different from the direction along which it originally aimed in flat space. As this line continues, it will eventually emerge into flat space once more, but in a different direction from the original line segment that started in flat

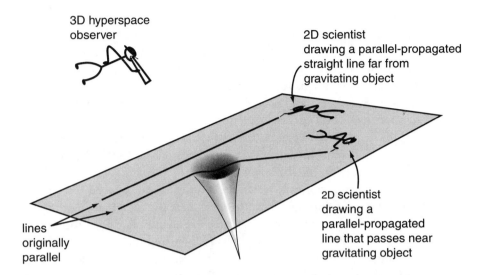

Figure 9.4. Two two-dimensional scientists draw parallel-propagated straight lines in their two-dimensional space. The lines begin parallel, but the one that responds to the curvature of the gravitating object will bend toward the center of curvature and emerge in a different direction. Both lines are legitimate straight lines in the two-dimensional space, even though one looks curved to a three-dimensional hyperspace observer.

space. This line is also a straight line in the two-dimensional curved space. From the superior three-dimensional position of the hyperspace observer the line looks curved. It is bent toward the center of the cone where the curvature is severe.

Looking from the point of view of the hyperspace observer is useful for perspective, but we must bear in mind that our reality is closer to that of the two-dimensional scientists. We must draw lines, do geometry, and figure out the curvature of space around gravitating objects as three-dimensional people in a three-dimensional space. We do not have the luxury of stepping out into some four-dimensional hyperspace and looking back to see how our space curves. We can determine that two initially parallel light rays passing by a star will diverge, just as the two scientists drawing the parallel-propagated lines in Figure 9.4 will determine a real divergence of initially parallel lines. The two-dimensional scientists cannot see the conical space around the gravitating object, as it is revealed to the hyperspace observer, but they can deduce its nature by doing careful geometry. They can, for instance, deduce that the radius of a circle in that part of space is long compared to its circumference.

We can explore the nature of space around a gravitating object a bit more. Think of an equilateral triangle composed of three straight lines surrounding the deepest point of the cone in Figure 9.4. Each line will look like an arc bowed outwards to a three-dimensional hyperspace observer. All observers will agree that the lines will not meet at 60-degree angles, and the sum of the interior angles will be greater than 180 degrees. How about parallel lines? Two lines drawn parallel initially will curve differently as they pass near the cone, and the one closer to the center will be bent more severely. The lines will not be parallel in the flat space to which they emerge. Lines drawn by parallel propagation will be the shortest distance between two points. A line that does not dip down in the cone must travel farther to reach a given point on the far side. Likewise, a line that goes too deeply within the cone will have wasted some motion and will have farther to climb out. There is a shortest distance between any two points, and the line that is shortest is straight, but there may be more than one straight line between two given points. Think of a line that misses the bottom of the cone narrowly to the left. It will be bent to the right. A line that misses the bottom to the right will be bent to the left. These two lines will cross. From the point of beginning to the point of intersection, there will be two straight lines.

All this is rather abstract, but it applies to Einstein's theory of gravity in general, not just in the vicinity of black holes. Think of the straight line that just encircles the neck of the cone and closes on itself, as shown in Figure 9.5. A straight line cannot do that in flat space, but the cone shows that it is not just possible but demanded of certain straight lines in the curved space. That closed curved straight line in curved space is an orbit! In Einstein's theory, orbits are not caused by the action of a gravitational force as they are in Newton's theory. For Einstein, the gravitating body causes a curvature in space – of which our cone is a representation – and orbiting bodies are moving with no force as straight as they can in that curved space. The Moon is moving as straight as it can in the curved space around the Earth, and the Earth is moving as straight as it can in the curved space around the Sun. For such problems as planetary orbits, both Newton's theory and Einstein's give virtually the same numerical results, despite the vastly different concepts on which they are based. That Einstein's theory explains everything that Newton did in the regime of weak gravity is one of the powers of the theory. In addition, Einstein's theory predicts the nature of black holes that Newton's is powerless to describe.

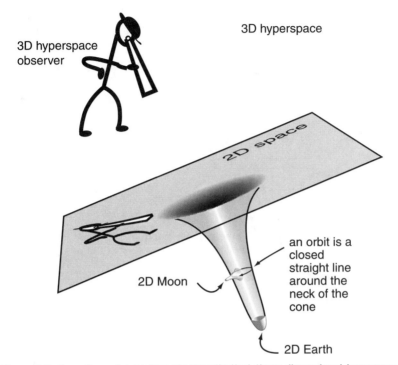

Figure 9.5. From the point of view of a hypothetical, three-dimensional, hyperspace observer, the space around the Earth would be a cone with the radius of a circle large compared to the corresponding circumference. The Moon moves as straight as it can by parallel propagating in the curved space around the Earth. In this conelike space, one set of straight lines consists of those that close on themselves around the neck of the cone. This is Einstein's version of an orbit. The Moon, in turn, causes space to be conelike in its immediate vicinity. This will cause rockets launched from the Earth to be deflected or to orbit even though they, also, are moving as straight as they can in the curved space. Note that the volumes of the Earth and Moon are reduced to areas in this two-dimensional representation.

Now, perhaps, you are prepared for the mind-bending exercise of attempting to picture the nature of curved space in its three-dimensional glory, with our toy two-dimensional cone as a guide. Figure 9.6 is an attempt to help do that. Draw a radial line out along the cone in the two-dimensional representation. At intervals, draw circles of constant radius, each with its own stretched-out radius. That will characterize the two-dimensional conelike surface as perceived by the three-dimensional hyperspace observer. What sort of three-dimensional curved space does the three-dimensional observer see in his own space? That's us! Imagine, if you can, rotating each of those

2D observer

3D hyperspace
observer

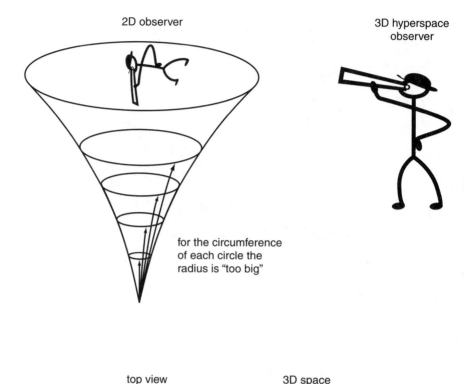

for the circumference
of each circle the
radius is "too big"

top view

3D space

space around
a black hole:
each inner
surface has
a smaller
circumference
and area, but
for each the
radius is
"too big"

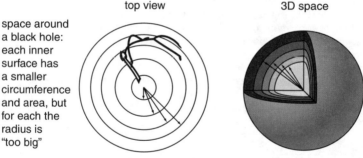

Figure 9.6. (Top) In the schematic two-dimensional curved space around a gravitating object, one can imagine circles of increasing radius and circumference. The circumference will always be smaller than 2π times the radius, and the discrepancy will
be largest for the innermost circles. Both the two-dimensional resident of the two-
dimensional space and the three-dimensional hyperspace observer will agree on that
general property, but the hyperspace observer can see the conelike space and the
reason for the large radius is obvious. (Bottom) If the nested circles of the top
diagram are rotated to map out a series of nested spheres, then one has a crude
representation of the space around a three-dimensional gravitating object. Each of
the spheres will have a circumference that is less than 2π times the radius. This is
impossible to represent in three-dimensional space (never mind on a flat sheet of
paper in this book!). A three-dimensional scientist could determine the curvature by
doing careful geometry but could never "see" the curvature of three dimensions.

circles in the two-dimensional space so that the swept-out locus of the rim of the circle is a two-dimensional sphere encompassing a three-dimensional volume. Now you have a set of nested spheres, but the distance from the center to the periphery of each sphere is "stretched out." The distance to the center of each sphere in the empty space around a gravitating object is larger than it would have been in flat space.

This exercise is an attempt to represent the curvature of the three-dimensional gravitating space. Neither the three-dimensional observer in Figure 9.6 nor we can directly perceive this curvature as a cone or anything else. For that, we would have to be a denizen of some fourth-dimensional hyperspace to look down on our three-dimensional space. We simply cannot do that. We can do careful three-dimensional geometry in the confines of our own three-dimensional space and work out the nature of the curvature of our space without ever being outside of it. If you were to measure the circumference of a given sphere around a gravitating object and then measure the distance to the center, you would find that the circumference was in every case less than 2π times the radius and that the smaller the sphere, the larger would be the discrepancy, just the property preserved in two dimensions and manifested in our cone representation. A three-dimensional scientist cannot, however, perceive where the extra length of the radius goes. All the scientist can or needs to know is that the radius is long compared to the circumference.

The important thing on which to concentrate is that such curvature exists in the space around the Earth, not just near a black hole. If you could draw a huge circle in the space around the Earth and then measure the radius of the circle, you would find that the radius was longer than you would expect if the space were flat. If you were to construct a triangle in the space around Earth consisting of three segments that are the shortest distances between the vertices, you would find that the angles added up to more than 180 degrees. All gravitating bodies curve the space around them! A black hole is only the most extreme example.

With this newfound perspective, let us return to the nature of black holes. Picture again a flat flexible sheet as a two-dimensional representation of flat, empty three-dimensional space. A star would cause a depression in the sheet. The star would be reduced to a two-dimensional spot of finite area (representing volume in the full three dimensions; check the Earth and Moon in Figure 9.5), and the depression representing curved space would extend beyond the star into the surrounding empty space. At no point within the star or beyond its surface is the curvature especially severe.

Suppose that the star were compacted to become a neutron star. This would be represented by making the spot smaller and the depression in the sheet much deeper. At rather large distances from the neutron star, the curvature of the sheet would be about the same. Near the neutron star, the walls of the depression will be nearly vertical (how one needs that three-dimensional, higher perspective to describe the goings-on!). As in the gravity of Newton, the strength of gravity depends on the distance to the center of the object. At the same relatively large distance, the gravity is the same. A neutron star has greater gravity than a normal star, not in the sense that it reaches out farther but in the sense that, because it is smaller in radius, one can approach much closer to the center of the gravitating star. A measure of the stronger gravity of the neutron star is the severity of the curvature of the flexible sheet at the bottom of the deep depression. The sheet would change directions rapidly at the bottom, a measure of the large curvature.

When a black hole forms, all the matter is crushed into the singularity. The mass of the star is no longer represented by an area but by a point. The flexible sheet is stretched to extremes. The curvature undergoes a discontinuity at the bottom of the cone. The sheet changes directions by 180 degrees in an infinitesimal length. One can go around the neck of the cone in an infinitesimal distance (see Figure 9.3). This is a representation of the infinite tidal forces that accompany a real singularity. Somewhere down inside the depression of the cone, a circle represents the location of the event horizon. To get the full effect, you should picture the space as an escalator moving rapidly inward, flowing down toward the singularity. To move outward, you have to run up the down escalator. At the event horizon, the escalator moves inward at the speed of light. Because you cannot run faster than the speed of light in the piece of space you occupy, you are dragged down to the singularity once you cross within the event horizon.

The singularity is a region of mystery, where our present laws of physics break down. That does not mean black holes cannot exist. Einstein's theory is still quite valid at the event horizon, which is the only part of a black hole anyone will ever observe and live to tell about. The British mathematician Roger Penrose has proved what is called the *singularity theorem*. This theorem says that once an event horizon forms by any means, some singularity must form. The theorem does not prove that *all* matter must fall into the singularity once a black hole forms, but that conclusion seems somehow inevitable.

5.2. Black Holes and the Nature of Time

Black holes cannot really be understood without a discussion of the nature of time in their vicinity. Like curved space, the flow of time is warped near and within a black hole. This makes temporal events difficult to picture in ordinary terms. One of the fundamental problems with a discussion of time in curved space is that everything depends on whose time you are discussing.

When two things are moving apart at a large relative velocity, the great Doppler shift means that all frequencies are observed to be lower. These frequencies include not only the frequency of light but also the tick of a clock, even the biological clock. Two people rocketing away from each other at great speeds will each see the other aging more slowly than they themselves are. In the case of large gravity, there is a related effect. To an observer who is not in a large gravitational field, a clock that sits deep within the gravitational pull of some compact star will be seen to run more slowly. A person orbiting around the compact star will be seen to age more slowly. The photons that climb out of the region of highly curved space and strong gravity require some time, so that the rate of arrival of the photons at a distant observer is slow. There is a long gap between the arrival of one photon and the next. Each photon carries information concerning the "age" of the object that emitted it. Because the photons take longer to get out, they arrive when the outside observer has aged considerably. The outside observer detects the photons and sees the object in the gravitational field as younger.

Consider two investigators. One volunteers to fall down a black hole, giving her life for science. The other, the project scientist, volunteers to remain at a safe distance and monitor the proceedings. The first volunteer falls straight down into the black hole and by her own watch and biological clock passes through the event horizon, is noodleized, and dies in a few seconds. The project scientist, watching through his telescope, sees the watch of the falling volunteer running ever more slowly, and the volunteer herself aging more slowly. As the falling volunteer approaches the event horizon, time stops flowing from the vantage point of the distant observer, and he never sees the falling volunteer cross the event horizon. The reason is that the last photon emitted by the volunteer before crossing the event horizon takes a very long time to reach the distant observer. The distant observer can, in principle, always see some photons from the falling person, no matter how long he waits. When those laggard photons finally arrive, the distant observer sees the falling volunteer before she crossed the event horizon.

In practice, the photons that arrive at distant times in the future are highly red-shifted and difficult to detect. In addition, the time between their individual arrivals is very long. Most of the time the distant observer sees absolutely nothing. Because of the large red shift and the delay between arrival of photons, the actual perception is that anything falling into the black hole turns black very rapidly.

The term "frozen star" was invented to describe the mathematical solution of Einstein's theory that corresponded to the result of the absolute collapse of a star. This term focused on the fact that a distant observer can never see the surface of the star fall through the event horizon. There is thus a suggestion that the surface of the star somehow lingers at the event horizon to be touched, and probed and explored. The term "black hole" was coined by John A. Wheeler in 1968 at a meeting in New York City on pulsars. Wheeler tried to come up with a graphic term to encourage his colleagues to contemplate even more extreme states of gravitational compaction than white dwarfs and neutron stars. The name "black hole" concentrates on the collapse and the fact that the star rapidly turns completely black and on the fact that, after collapse ensues, no part of the star can ever be recovered. If you tried to fly down and grab some of this frozen star, you would find that the surface receded from your grasp as your time became its time and you could see it fall once more.

The term "black hole" is much more pertinent to the real situation because it directs attention to the actual collapse and to the interior of the black hole. The case is difficult to prove, but there is a sense that the term "black hole" itself spurred some of the marvelous work that followed. With this new term and new mode of thinking came complete mathematical solutions of the interior of black holes, where people's minds can reach even if their bodies cannot.

6. Black Hole Evaporation: Hawking Radiation

As remarked earlier, Einstein's theory, for all its magnificence and success, is not complete. This theory is a so-called classical theory in that it incorporates none of the principles of the quantum theory. In Einstein's theory, as in Newton's, all motion and changes are smooth, and all positions can, in principle, be specified exactly. Einstein's theory is not compatible with our understanding of microscopic physics as described accurately by the quantum theory.

6.1. Quantum Event Horizons

The first successful attempt to include some of the principles of the quantum theory was done by the brilliant theoretical physicist from the University of Cambridge, Stephen Hawking. The process by which energy is converted into equal parts matter and antimatter is intrinsically a quantum mechanical process. Hawking's genius was to see how to add a little of the quantum process into the otherwise classical realm of Einstein's theory. He showed that the gravitational energy associated with the curved space in the vicinity of an event horizon will create particles and antiparticles. In principle, electrons and positrons, or even protons and antiprotons, could be generated. The easiest particle to make, however, is the photon because it has no mass (technically speaking, a photon and an antiphoton are one and the same thing).

According to the quantum theory, no position can be specified exactly. This applies equally well to the position of the event horizon around a black hole. Because of the intrinsic quantum mechanical nature of things, you cannot say definitely whether something is inside or outside the event horizon, only whether something is probably inside or outside the event horizon. The location of the event horizon is then fuzzy. When two photons are created in the vicinity of the event horizon, there is a probability – purely quantum mechanical in nature – that one photon will be inside the event horizon and will disappear down toward the singularity, and the other will be outside the event horizon and fly off to great distances where it can be detected. Hawking's great discovery was that black holes are not truly black. They shine with their own radiance generated from pure gravitational curvature!

6.2. A Two-Way Street

The physical implications of this discovery were immense and caused a wrenching turnabout in our view of black holes. The energy to create the radiation came from the gravitational field, but the gravitational field came from the mass of the matter that had collapsed to make the black hole. When the photons carry off energy, the energy of the black hole must decline. This can only happen if the mass of the black hole declines as well. As black holes emit Hawking radiation, they are shining away their very mass! Black holes are not completely one-way affairs after all. Even though it is still true that tidal forces will tear an object beyond recognition as it falls into the singularity, the mass is not gone forever. It will emerge later in the form of the Hawking radiation to permeate the Universe. A black hole is just nature's way of turning all that bothersome matter into pure random radiation. We will see that nature has yet other tricks with the same fate in mind. Gather ye rosebuds while ye may, a photon yet ye'll be!

Hawking discovered that the black hole radiation does not come out in an arbitrary fashion. The spectrum of the radiation corresponds exactly to a single temperature, when it might have been some odd, nonthermal shape. The temperature is determined in turn by the mass of the black hole. The variation with mass is inverse so that a massive black hole has a low temperature, and a low-mass black hole has a higher temperature. For a black hole of stellar mass, the temperature is very low. Little radiation could be emitted in a time as short as the age of the Universe, and so the radiation is of little practical importance. Our standard picture of black holes as gaping one-way maws holds true.

6.3. Mini-Black Holes

If the mass of the black hole should be less than that of an average asteroid, however, the situation is markedly different. Such small black holes would be very hot and would radiate prodigious amounts of radiation. As these small black holes radiate, their mass shrinks so they get hotter and radiate even faster. The process runs away faster and faster. In less than the age of the Universe, such small black holes could evaporate completely! The final stages of this process are so accelerated that the last energy would emerge in an explosion of high-energy gamma rays.

These so-called *mini-black holes* could not be created in the collapse of an ordinary star. They might have arisen in the turbulence that may have marked the original state of the big bang. If this were the case, there could be swarms of mini-black holes in the Universe, some of which would be explosively evaporating at any time. The properties of such explosions have been worked out theoretically, and the radiation has been sought, but so far unsuccessfully.

6.4. White Holes

One can imagine (mathematically) the reverse of a black hole, or a *white hole*. A white hole is obtained by running time backward compared to the flow of events for a black hole. For a black hole, one starts with ordinary space. A star collapses to make a black hole, and then you have a black hole forever, gobbling up matter, but releasing nothing (forgetting for the moment Hawking radiation). Now run the movie backward in time. One must start with a white hole that has existed since the beginning of the Universe, spewing forth matter but swallowing nothing. At some time, the "last stuff" pours forth, and one is left with empty, flat space.

Black holes are regarded seriously because we can predict that they might well occur in the course of stellar evolution and because we think we have found them, as Chapter 10 will show. From the properties of known stars, the properties of the resulting black holes can be predicted. White holes are not regarded on the same footing because they must exist since the beginning of time. Their properties cannot be predicted because we cannot predict the beginning of the Universe. White holes could have any property – large mass or small. Because we cannot predict their properties, white holes have no firm place in the realm of ordinary pragmatic physics.

Hawking's discoveries may have been a first step toward putting the notion of white holes on a firmer basis. Hawking has blurred the distinction between white holes and black holes by introducing quantum mechanical properties to the event horizon. Now we see that a black hole can emit radiation, a property previously reserved for white holes. Likewise, a white hole should be able to swallow radiation. Hawking has argued that for very small objects the distinction between white holes and black holes may disappear.

7. Fundamental Properties of Black Holes

For all their exotic nature and the complexity of the theory that treats them, black holes can have only three fundamental intrinsic properties. These properties are their mass, their spin or angular momentum, and their electrical charge. These properties are distinguished because they can be measured from outside the black hole and, therefore, determined by ordinary techniques. The mass can be determined by putting an object in orbit around the black hole and seeing how fast it moves. The charge can be determined by holding a test charge and detecting the force of attraction or repulsion from the hole. In practice, one expects real black holes to be electrically neutral because they should rapidly attract enough opposite charge from their surroundings to neutralize any charge that might build up. Measurement of the spin of a black hole is a more subtle process. As the black hole rotates, it drags the nearby space around with it. This dragging can be measured in principle, like the currents in the ocean. Once the mass, spin, and charge of a black hole are known, all its other intrinsic properties are set. For instance, for a noncharged, nonspinning black hole, the size given by the radius of the event horizon is strictly proportional to the mass. The temperature of the Hawking radiation varies inversely with the mass. Other properties that a black hole might have, but cannot, are mountains like the Earth or sunspots and flares like a star. On a more fundamental level, black holes cannot have the property of a lepton number or a baryon number. The forces associated with leptons and baryons are short range and cannot extend outside the event horizon where they can be measured. Black holes do not so much violate the laws of conservation of lepton and baryon number as transcend them. In the realm of black holes, these fundamental physical laws of ordinary space are irrelevant. John A. Wheeler has coined an aphorism to describe this raw simplicity of black holes – he says "black holes have no hair."

To illustrate the power of this notion, consider two compact stars. Let one be made of neutrons, an ordinary neutron star. Let the other be made of antineutrons, an antineutron star! If these two stars were to collide, the neutrons and antineutrons would annihilate to produce pure energy and an explosion of unprecedented proportions. Suppose, however, we dump a few too many neutrons on the first star and it collapses into a black hole. Then we add some antineutrons to the second star so that it, too, collapses to make a black hole. Do we now have a black hole and an antiblack hole? No, we have two identical black holes because the black holes transcend the law of baryon (neutron and antineutron) number. If the two black holes combine, the result is not an explosion but one larger black hole. The form of mass that originally collapsed to make a black hole becomes irrelevant after it has passed through the event horizon. Then only the total mass counts. While he was warming up, Stephen Hawking presented to the world the laws by which black holes combine to make larger ones, an exercise that alone would have assured his reputation as a brilliant physicist.

8. Inside Black Holes

Just because black holes have only three fundamental properties does not mean that their nature, which derives entirely from specifying the values of those three properties, is not complex. Apart from quantum effects, the exterior of a black hole, the event horizon, is a model of simplicity: smooth, perfect, and unperturbed. The insides, however, as exposed by the powerful techniques of mathematics, are a wonder such as to strain one's credibility to the limits.

8.1. Timelike Space

When we discussed the oddities of the flow of time near black holes (Section 5.2), we omitted the oddest twist of all. This aspect can never be observed directly, but it is the real factor that accounts for the existence of the event horizon that blocks our view. Inside the event horizon, space takes on the aspects of time (cf. Figure 9.1). No matter how rockets are fired or forces applied, any object must move inward toward the singularity (or outward, if we are dealing with a white hole) as it ages. There is no choice in the matter, just as you have no choice in the matter of your aging from eighteen to thirty-one. The same principle that drags you on into old age drags an object within the event horizon ever closer to the singularity. Within the event horizon, space is no longer the entity in which you can move around in three dimensions with impunity. There is only one direction, inward. The one-way nature of this space is intimately related to the one-way nature of time. Inside a black hole space is timelike! The timelike nature of space is the reason that everything goes inward inside a black hole, and nothing can get out. It is the reason black holes are black.

8.2. Schwarzschild Black Holes

The simplest black hole is one with mass, but no charge or spin. This kind is called a *Schwarzschild black hole* after the physicist who first gave a mathematical description of such a beast, shortly after Einstein presented his general theory of relativity. There is a poetry to this name that is rendered as black shield from the German. This was the type of black hole illustrated schematically in Figure 9.1.

For a Schwarzschild black hole, the event horizon coincides exactly with what is called the *surface of infinite red shift*. A photon emitted from this surface will have an infinitely long wavelength by the time it escapes to great distances. The event horizon is round for a Schwarzschild black hole, and the singularity is a point at the center of the black hole.

Mathematical investigations have shown that even the lowly Schwarzschild black hole is not so simple. In the idealized case, where one assumes that all the mass is confined to the singularity and that a vacuum exists everywhere else, a black hole is really twain, two equal geometries sharing the same singularity. Each black hole has its own Universe of empty flat space. These two Universes exist at the same instant but in different places. When moving at less than the speed of light, one cannot travel from one to the other but will instead fall into the singularity if passage between them is attempted. This idealized mathematical description does not apply to a black hole that has formed from the collapse of a star. Then the matter of the star introduces other changes in the geometry and curvature of space that are, as yet, too complicated for anyone to have been able to calculate. The "other Universe" is undoubtedly just a mathematical fiction, but it gives a portent of the richness to come.

8.3. Kerr Black Holes

One has only to introduce some rotation to the black hole to complicate affairs in the most interesting fashion. The first basic mathematical solution corresponding to rotating black holes was discovered by the New Zealand physicist Roy Kerr in 1963. Subsequently, the complete solution of the interior of a rotating black hole was worked out by others, but these black holes are still referred to as *Kerr black holes* to distinguish them from Schwarzschild black holes.

If a black hole rotates rapidly enough, the event horizon disappears completely. In this case, one could look directly into the fearsome maw of the singularity. Such a beast is known as a *naked singularity*, a singularity unclothed by an event horizon. There is no formal proof as yet, but there is a strong belief that no black hole can rotate fast enough to create a naked singularity. Certainly any star that rotated so fast would fling itself apart before it could collapse to make a black hole. Firing matter into a black hole tangentially would spin it up. Calculations show, however, that as the black hole nears the limit where the last veil might be dropped, gravitational radiation will become so intense as to carry away any increment in rotational energy. Perhaps there is some way to create a naked singularity, but it seems very difficult. Many researchers have adopted the as yet unproven doctrine that naked singularities cannot exist in the real world of astrophysics. This doctrine that nature denies freedom of expression to unclothed singularities is known informally as "cosmic censorship." Stephen Hawking, a firm believer in cosmic censorship, bet Kip Thorne of Caltech that naked singularities cannot exist. He paid off on the bet when the carefully designed computer models of Matt Choptuik yielded naked singularities. No one has yet found one in their backyard.

Real rotating black holes may have matter swarming around inside the event horizon that will substantially alter the geometry of the inner reaches. The best we can do is to follow the mathematician's description of the idealized case where, once again, the assumption is made that all mass is confined to the singularity, and that all the rest of space is pure vacuum. Welcome to Wonderland, Alice!

The first thing one discovers in the study of rotating black holes is that the singularity is not a point but a ring! One can imagine an intrepid explorer plunging through the center of the ring, avoiding the infinite tidal forces of the singularity itself. Retreating now to the outside, we find that for a rotating black hole the surface of infinite red shift separates from the event horizon. Both surfaces are oblate, flung out around the equator by centrifugal forces, but the surface of infinite red shift is more extended. There is a finite distance between the surface of infinite red shift and the event horizon at the equator. At the poles of the rotation axis, the two surfaces are still contiguous.

The surface of infinite red shift has another property. It is also the *stationary limit* with respect to sideways motion. The rotation of the black hole drags the local space around in the same sense as the hole rotates. The effect is stronger the closer one is to the black hole. At a moderate distance, one could fire rockets and overcome the effect in order to hover in one place. This requires some effort like swimming upstream or walking up the down escalator. At the stationary limit, all efforts to remain still are fruitless. To resist moving around in the same sense as the black hole spins, one would have to fly backward in the local space faster than the speed of light. Inside the stationary limit, all material objects, including photons of light, are forced to rotate with the hole.

On the other hand, because the surface of infinite red shift is removed from the event horizon at the equator, one can, in principle (ignoring the huge tidal forces), fly inside the surface of infinite red shift and return. This can be done by moving with the rotation of the black hole, the path of least resistance. Some paths lead into the event horizon, and there will be no return; however, with a rotating black hole, the option exists to emerge from within the surface of infinite red shift.

The region between the surface of infinite red shift and the event horizon is called the *ergosphere*. This phrase was coined by Roger Penrose (of the singularity theorem) who investigated its properties. It derives from the Greek word ergo, meaning work or energy. Penrose found that, under proper circumstances, energy could be extracted from the black hole. If one of a pair of particles is fired down the hole in a counterrotating sense from within the ergosphere, the recoil will throw the other particle out with more energy than both particles had originally, including their mass energy, $E = mc^2$. You do not get something for nothing. In this case, the excess energy in the ejected particle comes from the rotational energy of the black hole. After the particle is ejected, the black hole will be rotating less rapidly.

There is some question as to whether this *Penrose process* for tapping the energy of a rotating black hole can be of real astrophysical interest. The problem is that a considerable investment of energy must be made in firing the first particle into the event horizon in the proper fashion. A puny nuclear explosion would be far from sufficient; the particle must be moving at nearly the speed of light. Such reactions with massive particles may not occur spontaneously in nature with any reasonable probability. On the other hand, photons are already moving at the speed of light. There have been discussions of Penrose processes operating to swallow some photons and eject others at high energy. This process is also driven by the rotational energy of the black hole and is termed *superradiance*. There is some speculation that the gamma rays seen from quasars could be produced in this way, starting with photons in the more conventional X-ray or ultra-violet range that are produced in the inner edges of a hot accretion disk.

Let us now journey into the event horizon. As we pass within, we come to a region of timelike space in which we must move inward as we age. There is a crucial difference in the rotating case, however, for there is an inner boundary to this timelike region. At this inner boundary is another event horizon, which prevents a return to the space beyond. Within this second event horizon is a region of normal, if highly curved, space. This event horizon prevents a return to the timelike space, rather than preventing a return to normal space.

Within this inner volume of normal space is another surface of infinite red shift, but because one can move in and out of such a surface if appropriate moves are taken, it has no direct consequence. Around the equator of this inner surface of infinite red shift is the line we devoutly wish to avoid. That equatorial line is the location of the ring-shaped singularity. If we stumble against that, we are doomed by the infinite tidal forces.

The special property of this inner region of normal space is that we could elect to stay here forever. By careful choice of movement, we can orbit around and never strike the singularity itself. This is very different from the case for a nonrotating black hole. There the time-like space leads inexorably to the singularity.

Other options await if we continue our imaginary journey within the spinning black hole. At the same place, but in the future, there is a similar space-time structure. Here, however, the sense of the event horizons and timelike space are reversed. As one flies about, one could in principle elect to head outward, passing through an event horizon into a region of outgoing timelike space. This would be bounded by an outer event horizon, and beyond that would be an ergosphere, a surface of infinite red shift, and finally free space. Formally, mathematically, this is not the space from which we entered, but another, separate Universe. The mathematical solution shows that in this new Universe there will be another in-going black hole like the original one we entered, so one can plunge down again and come out in yet a third Universe. The idealized mathematical solution we are exploring has an infinite number of Universes, all connected by rotating black holes!

Let us return to the central regions of the rotating black hole. We found there a more or less spherical region of normal space inside of which lay the ring singularity. Watch carefully now, Alice! The plane of the ring singularity divides the volume into two halves. You can maneuver from the top half, out through the inner surface of infinite red shift, and back in, to come to the bottom half. Alternatively, you could elect to plunge straight through the hole in the middle of the ring. In so doing, you would come to *a* bottom half, but not the one accessed by going out and around the ring. If from this new lower half you went out and around, you would be in a top half, but again not the one from which you started. The space through the ring is not the space you get to by going around the ring. If this is not passing through the looking glass, what is? You can imagine looking down through the ring and seeing another creature, perhaps a puce-colored eight-legged cat. If you go out around the singularity and look, you will not see the creature. Its space is only through the ring, not behind it.

If you join the creature through the ring, you can seek, in the future, a set of outgoing event horizons. These will again lead to an outer, flat Universe, which is none of the ones we have discussed previously. As you leave this black hole, you will feel it pushing you. Unlike the others we have explored, this out-going solution that exists through the ring antigravitates!

Having entertained ourselves thus, we must return to more sober reality. We do not diminish the wonder of the tale to point out again that what has just been described is an idealized mathematical solution. It is a marvelous, exact solution to the full set of equations describing general relativity. Nevertheless, a crucial assumption has been made in order to solve the equations at all. The assumption is that there is no mass anywhere except in the singularity. The presence of any matter or energy within the first set of event horizons would cause a change in the curvature and geometry, and the wonderful world of multiple Universes would probably vanish. The solution to the equations with even a little matter present throughout the volume would not contain any of the extra spaces, in the future or through the ring. Even the presence of an explorer such as we imagined ourselves to be could change the whole situation.

Some research has been done to see what happens to the mathematical solution if the tiniest bit of extra matter is added inside the black hole. There is a strong suggestion that the whole geometry would begin to rattle and shake with the resultant generation of an intense flux of gravitational radiation. This radiation alone would alter the physical and mathematical situation, to eliminate the reality of the extra spaces and Universes. At the very least, in the real Universe, photons of light will continue to flood down the black hole. As they plummet in, they are blue-shifted and attain incredible energies. This energy will build up at the event horizon in what has been termed a *blue sheet*. This sheet of energy would warp the geometry and wipe out any of the multiply-connected interior geometry.

The mathematical "vacuum" solution to the Kerr black hole is a marvelous, mind-stretching exercise. It probably has nothing to do with the guts of a real star-born black hole, rotating or not. On the other hand, the reality is fantastic enough as we shall see in Chapter 10, and the mystery of the singularity remains.

10

Black Holes in Fact

Exploring the Reality

1. The Search for Black Holes

Black holes, those made from stars, are really black! How can we hope to find them if they do exist? Some solitary massive stars may collapse to make isolated black holes drifting through the emptiness of space. There could be very many of these black holes. Estimates based on the number of massive stars that have died in the history of our Galaxy range from one to a hundred million black holes. The simple fact is that, until a space probe stumbles into one, we are likely never to detect this class of isolated, single black holes. We will certainly never see the black hole itself in any circumstances because no light emerges from it. Our only chance to detect the presence of a black hole is to find a situation where mass is plunging down a black hole, heats, and radiates. We can hope to detect the halo of radiation from such an accreting black hole even if we never see the black hole itself. Black holes are so strange and so significant that the standard of proof must be exceedingly high. As we will see, the evidence is very strong, but still largely circumstantial.

Many astronomers search for giant black holes in the centers of galaxies. The evidence for those black holes has become rather strong in the last few years, but most of the evidence still involves matter moving far beyond the event horizon, and we know very little about the configuration of the accreting matter. There is no question that there are concentrations of gravitating mass in the centers of galaxies, including our own, that contain millions if not billions of solar masses, are small, and are not radiating anything like an equivalent amount of star light. One idea is that they could be a cluster of compact stars, neutron stars, or stellar mass black holes, but the theory of such swarms of objects says they should quickly collide and merge and make one large black hole. With some theoretical underpinning and compelling circumstantial evidence, the argument for these giant black holes is rather convincing. There are clues from the X-rays from some galactic cores that the space near the very center has just the character you would expect for that around a rotating, supermassive, Kerr black hole. More evidence of this kind may remove any ambiguity.

Another excellent hunting ground for black holes has proved to be in binary star systems where mass transfer can feed the accretion and produce X-rays in the high gravity of a stellar-mass black hole. Here also the case has become very strong that we are observing black holes. This facet of black hole research is closely connected to the topics covered in this book, so this story is worth telling in more detail.

Over thirty strong X-ray sources have been established to be in binary systems. Of these systems, about a dozen have some determination of the mass of the X-ray source itself. In most cases, the mass is in the range of one to two times the mass of the Sun. These are probably neutron stars. In some cases, pulsations are observed, and the case for rotating, magnetized neutron stars is clearly established. One should perhaps bear in mind, however, that, although a neutron star cannot have a large mass, there is no reason in principle why a black hole could not have a modest mass, particularly if it formed by adding a bit too much mass to a neutron star. We still have no unambiguous way of determining that we have a black hole with a mass less than the maximum mass of a neutron star, although there are some ideas for how to do this.

In the case of a black hole, there is no question of radiation from the surface of the object because there is no matter, only the ephemeral event horizon. All the X-rays must come from matter in the accretion flow. Within about three times the radius of the event horizon of a black hole, the gravity is so strong that the matter cannot spiral in a disk but must plunge headlong into the hole. In this state, the matter radiates much less because it is not subject to the friction of the accretion disk. In addition, the radiation emitted from this region is highly red-shifted, so it is difficult to detect with X-ray devices. Any X-rays detected from an accreting black hole will come from a halo in the disk, inside which there is only blackness. This particular way in which X-rays are emitted may prove sufficiently different from the X-ray emission mechanisms for neutron stars that black holes can be unambiguously identified, independent of their mass. For now, the story is a bit less certain.

2. Cygnus X-1

One of the first binary X-ray sources discovered is a candidate black hole system. This object was the first X-ray source discovered by the *Uhuru* satellite in the direction of the constellation Cygnus. Soon after its discovery, astronomers were describing Cygnus X-1 as a possible black hole. Absolute proof escapes us, but the net of circumstantial evidence has grown ever tighter. Cygnus X-1 is probably a black hole.

The chain of arguments proceeds like this. The fact that Cygnus X-1 emits a strong flux of energetic X-rays at all argues that it is a compact object with a large gravitational field. It could be a white dwarf, a neutron star, or a black hole. The intensity of the X-rays argues against the white dwarf possibility. Added evidence against a white dwarf is that the X-rays from Cygnus X-1 flicker on a time scale of milliseconds. We can use an argument based on how far light can go in a given time to say that the object must be smaller than the distance light can travel in 0.001 second. That distance is 30 kilometers, consistent with a neutron star or a black hole, but too small to be a white dwarf. A white dwarf would be too large and sluggish to vary rapidly. The conclusion that Cygnus X-1 is not a white dwarf, never mind an ordinary star, seems quite sound.

This leaves us with a neutron star or a black hole as the necessary object. There may be a foolproof way to tell the difference from the nature of the X-ray emission alone, but that argument is still under development and is difficult to apply cleanly to Cygnus X-1. Many feel that the millisecond fluctuations are themselves evidence of the nature of a black hole, but that has not been proven. The lack of regular pulsations is not sufficient because the object could be a slowly rotating or unmagnetized neutron star that could not produce detectable pulses. The only way we know to distinguish between a neutron star and a black hole is to argue that a black hole can exceed two or three solar masses, and, as discussed in Chapter 8, a neutron star cannot.

Careful study of the Cygnus X-1 system, both the X-ray source and its companion massive star, shows that the companion has a mass of about 30 solar masses, and the X-ray source, a mass of about 10 solar masses. The latter is too much to be either a white dwarf or a neutron star. By a process of elimination, the reasonable conclusion seems to be that Cygnus X-1 is a black hole.

The presumption behind this chain of reasoning is that the massive star transfers mass to the black hole, and the infalling matter emits X-rays before it plunges into the black hole, but all we really know for Cygnus X-1 is that a 10 solar mass "thing" is emitting X-rays. As an example, let us consider a way in which nature might be playing a trick on us. We know that triple-star systems are present in the Galaxy. We noted in Chapter 3 that the nearest star, Alpha Centauri, is in a triple system. Suppose that Cygnus X-1 consists of a neutron star of 1 solar mass orbiting an ordinary star of 9 solar masses, and that the pair of them are orbiting another ordinary star of 30 solar masses. If the 9-solar-mass star transfers mass to the neutron star causing the emission of X-rays, then we will have an X-ray source with total mass of 10 solar masses orbiting a 30-solar-mass star, just as the observations demand, yet there would be no black hole. This picture is unlikely, but not entirely impossible. The reason we can consider it at all is that the 30-solar-mass star would be considerably brighter than the 9-solar-mass star, so the latter could be lost in the glare. Attempts have been made to detect such a masquerading companion by searching for faint spectral lines that would shift around among the spectral lines of the brighter star as the Doppler shift responds to the orbital motion. No hint of such a secondary star has been forthcoming. It probably is not there, but a tiny doubt will always linger.

The massive companion to the X-ray source in Cygnus X-1 is blowing a stellar wind, as such stars do. The picture adopted for Cygnus X-1 is that the gravity of the black hole traps part of the wind. That matter then swirls into an accretion disk. The matter then spirals down, and the friction heats the gas to temperatures where the matter radiates X-rays. The companion is transferring mass at a sufficiently slow rate that it seems unlikely that the black hole in Cygnus X-1 could have started as a neutron star and then collapsed to a black hole and subsequently grown to its present mass before the companion died. The presumption is that the black hole formed directly by the collapse of a 10-solar-mass object.

It does not follow that the black hole arose from a star whose initial mass was only 10 solar masses. A more likely prospect is that the progenitor star had a mass of around 35 solar masses. The other star, the normal companion that still exists, probably had about the same mass we see now, around 30 solar masses. Stars of 30–35 solar masses develop helium cores of about one-third their original mass. The originally more massive star thus probably grew a helium core of about 10 solar masses as it burned up the hydrogen in its center. At the same time, the star probably lost a great deal of mass due to its own stellar wind. The most likely time for this is when the originally more massive star finally exhausted its central reserve of hydrogen and began to become a red giant. At this time, any mass remaining above the helium core probably flowed out of the binary system or onto the companion star. During this episode, the companion could have lost some mass to a wind and gained some from the more massive star, so it did not change appreciably.

Even though it has lost its hydrogen blanket, the now bare 10-solar-mass core of the first star is so massive that it is supported by the thermal pressure and continues to evolve with regulated nuclear burning. The core presumably burns a series of nuclear fuels until it forms an iron core. This core collapses, but instead of producing the explosion of a supernova, a black hole forms. All the matter in the core rains down through the event horizon. The net effect is that the 10-solar-mass black hole did not come from a 10-solar-mass star but more likely from one originally with somewhat more than 30 solar masses. The corollary implication is that this star did not explode but left a black hole instead. One is invited to think that all stars in this mass range, greater than 30 solar masses, leave black holes. A possible problem with this reasoning is that the very fact that the star was in close orbit with a massive companion may have altered the evolution in a way we do not understand. As we discussed in Chapter 6, there is little direct evidence concerning the end point of massive stars of a given initial mass. In any case, a common presumption is that stars of about 30 solar masses must explode to provide the heavy elements. Clues that stars of this mass make black holes means that there is no strong evidence to support this presumption.

3. Other Suspects

Further observations showed that there are binary systems emitting X-rays that provide even better evidence for black holes than the famous Cygnus X-1.

One of these systems is LMC X-3. This object is the third X-ray source discovered in the nearby galaxy, the Large Magellanic Cloud, which also played host to Supernova 1987A. LMC X-3 is similar to Cygnus X-1 in that the X-ray source seems, from a study of orbital parameters, to have a mass of about 10 solar masses, and hence to be too massive to be a neutron star. In this case, however, the companion star is only about 10 solar masses as well. This means that it is much more difficult to hide a third star in the glare of the ordinary star than in the case of the more massive, and brighter, companion in the Cygnus X-1 system. A three-body system with a neutron star orbiting an undetected normal star, with both orbiting the observed normal star, would be untenable. There would be obvious evidence of the third star. LMC X-3 may thus be a better candidate for a black hole than Cygnus X-1 because one cannot resort to the dodge of hiding some other source of mass and gravity in the system.

There is, however, a system in our Galaxy that is an even better candidate for containing a black hole in orbit. That is the system with the boring moniker AO620-00, named for its directional location in the Galaxy. This system seems to have a 5-solar-mass black hole orbiting a normal star that is not massive at all but about one-half the mass of the Sun. It is not clear how a star with original mass of about 30 solar masses, that could have a core of about 10 solar masses, which in turn could collapse to make a black hole, would come to have such a wimpy companion. Usually massive stars seem to hang out with one another. On the other hand, nature may be tricking us here. If every 30-solar-mass star had a 0.5 solar-mass companion, the dinky star would be lost in the glare, and we would never know it. Nature may form stars in this way much more frequently than we realize, or there may be something else going on that is special to black hole systems. One suggestion is that the little companion star forms from the matter spun off the star that forms the black hole. In any case, the small-mass, dim companion means that it is virtually impossible to hide another star in the system to trick us into thinking that an X-ray-emitting neutron star had a higher mass, therefore masquerading as a black hole.

Another argument adds to the case. AO620-00 underwent at least two outbursts that produced an excess light output, one in 1917 and one in 1975. The 1975 eruption produced a corresponding detected burst in X-rays. These bursts lasted for about a month and, in the optical at least, are rather reminiscent of dwarf nova outbursts. Models of the behavior of accretion disks around black holes reproduce the properties of the optical and X-ray bursts with the same kind of physics that works for dwarf novae, as discussed in Chapters 4 and 5. The accretion disk collects matter until it undergoes an instability that dumps matter into the black hole at a greater rate, resulting in the outburst.

The arguments are still circumstantial. What we know is that AO620-00 contains an orbiting object with a large mass that emits X-rays but virtually no optical light. Nevertheless, it is very difficult to see how AO620-00 could be anything but a black hole.

There is a bit of a tendency to cry "black hole" whenever a strange new astrophysical phenomenon involving high energies turns up. That is one reason most astronomers are trying to be as conservative as possible about concluding that Cygnus X-1, LMC X-3, and AO620-00 are black holes. There is another danger: there are other black holes out there, and we are being too conservative to face the facts. The last few years have revealed that the Galaxy is full of systems like AO620-00.

4. Black Hole X-Ray Novae

One way to beef up our confidence that Cygnus X-1, LMC X-3, and AO620-00 are black holes is to find others. There is safety in numbers. The problem is that the combination is rare wherein a massive star makes a black hole, and we catch a comparably massive companion as it is transferring mass, but before the companion also dies. Only about one such pair should exist in the Galaxy at any one time. We may have discovered that one rare event in Cygnus X-1. It is possible that LMC X-3 is the only currently active black hole with a massive companion in that smaller galaxy, just as Cygnus X-1 may have that single merit in our Galaxy. The formation of black holes is associated with massive stars, and Cygnus X-1, the grand-daddy of black hole candidates, has a massive companion. The feeling lingered for a long time that all black hole binaries, if they existed, would resemble Cygnus X-1. In the last decade or so, we have learned that the Galaxy is full of binary black hole candidates, but, like AO620-00, they are wonderfully and surprisingly different from Cygnus X-1. These systems are even better candidates for black holes than the venerable Cygnus X-1, and they present better laboratories to explore the astrophysics of black holes.

Two basic characteristics distinguish the new class of black hole candidates, of which AO620-00 is the prototype. They show a distinct transient behavior, and they have low-mass, relatively dim companions. These systems maintain a quiescent state for decades and then erupt in a sudden burst of energy. The energy output appears throughout the range of electromagnetic waves from radio to gamma rays. There is especially interesting behavior in the soft and hard X-ray bands. The outbursts last for about a year, and then the system fades to quiescence again. In the quiescent state, the only evidence of the system is the small-mass companion. Without an eruption to draw the attention of astronomers, these stars are lost among the billions of similar stars in the Galaxy. Without the ability to detect the associated high-energy emission in X-rays and gamma rays, even the outburst may pass without special notice. Such eruptions may have been mistaken for classical novae in the past.

When AO620-00 underwent an outburst in 1917, before the invention of X-ray astronomy, it was taken for an ordinary nova. AO620-00 had a dramatic X-ray outburst in 1975, but it was several years later before evidence came in that it might harbor a black hole. Only relatively recently has the realization dawned that the Galaxy contains many of these systems. The coverage of the sky with satellites that can monitor X-ray outbursts has been fairly thorough for the last decade. The result is that astronomers have discovered X-ray novae that are black hole candidates at the rate of about one per year in the Galaxy for the last 10 years. Because these systems sit quietly undetected for perhaps 50 years for every year they are in outburst, then every one outburst may represent 50 sleeping systems. Our vigilance in watching the Galaxy is not perfect, and gas and dust could obscure some events. Allowing for such problems, one can guess that there could be 100 to 1,000 such black hole systems in the Galaxy. Thus they vastly outnumber systems like Cygnus X-1.

One of the principal goals in the study of these erupting systems is to find proof that they contain black holes, not neutron stars or some other configuration that can mimic the circumstantial evidence for a black hole. Currently the most reliable way to establish a black hole candidate is to show that the compact object in a binary system has too much mass to be a neutron star.

Five or six black hole novae are excellent black hole candidates. These systems have at least a firm lower limit to the mass of the object emitting the X-rays that rules out a neutron star. Among these are AO620-00, V404 Cygni and Nova Muscae 1991. V404 Cygni is currently the best candidate for a black hole in a binary system. Many careful observations reveal that the mass of the compact star is about 12 solar masses, far more than is possible for a neutron star. Approximately another two dozen systems are good black hole candidates based on the similarity of their optical and X-ray outburst behavior to the temporal and spectral behavior of the best established candidates.

In most of the black hole X-ray novae, the companion has a small mass. The companion stars are dim and hence difficult or impossible to detect even when the system is at minimum light. In the systems where information is available about the mass of the compact object, there is also information about the mass of the companion. In AO620–00 and Nova Muscae 1991, the normal star companion is substantially less than 1 solar mass. V404 Cygni is somewhat a special case. The companion has evolved past the main sequence stage, but even then the remainder of the star has a mass of only about 4 solar masses. For most of the systems, the companions are low-mass, low-luminosity stars, with a mass considerably less than the mass of the putative black hole. There is no question of a third star masquerading in any of these systems, adding mass that would be mistakenly attributed to the compact object.

5. The Nature of the Outburst

To obtain a basic understanding of the behavior of these systems one of the most important questions to address is the reason for the outburst. The most promising model for the basic outburst is an instability not directly associated with either the black hole or the companion star, but within the accretion disk that passes matter between them. The companion star provides the reservoir of mass. If the mass flows too slowly from the companion, the accretion disk cannot remain in a hot, ionized state, and a steady rate of flow is not possible. These systems must undergo accretion disk outbursts similar to those in dwarf novae and some neutron star binary systems, as discussed in Chapters 5 and 8. In the simplest picture, the disk flares to make excess optical and X-ray radiation and then goes back into storage mode, accepting matter from the companion, but passing very little through itself and down the black hole. The disk emits little optical light and virtually no X-rays. The main thing observable in this state would be the companion star and perhaps the spot on the edge of the disk where matter rains in from the companion. The disk could develop a very hot, nearly spherical inner region, as discussed in Chapter 4 (Figure 4.6), which would alter this simple picture and give another source of luminosity in the "off" state. We will return to this topic in the next section.

This physical process of the disk instability does not depend on the exact nature of the compact object or of the star providing the mass. It can happen to accretion disks surrounding white dwarfs and neutron stars as well as black holes. The majority of the X-ray novae that display this outburst behavior show no explicit evidence for neutron stars and remain black hole candidates.

The disk outburst model can account for the decade-long periods of quiescence, which are set by the time for matter to collect or ooze inward in the cold, low-viscosity disk. The rapid rise time of days can be associated with the time scale for heating waves to propagate in the inner disk. The year-long decline is governed by the more rapid viscous evolution in the hot state and the time for the cooling wave to propagate through the disk.

There are some explicit tests of this picture. The model predicts that in quiescence, the mass transfer rate as determined from the luminosity of the "hot spot" where the accretion stream collides with the disk should be far greater than the flow into the black hole as determined from the X-ray luminosity produced in the inner disk. These basic predictions are borne out by optical and ultraviolet observations of AO620–00 from the *Hubble Space Telescope* and X-ray observations with the German *ROSAT* satellite. Other confirming evidence comes from the lack of helium emission lines. If the inner regions generated X-rays, the X-rays would excite the gas to produce fluorescent emission lines. The lack of those spectral features means that there cannot be many X-rays and hence little mass flow in the inner disk. These observations seem to show that the disk is storing matter.

One objection to the model is that the disk does not seem to cool in the decline phase as much as predicted. This may be due to the formation of a hot "corona" around the disk, much like the corona that surrounds the Sun. In that case, the observed surface temperature does not reflect the temperature of the body of the disk that the models predict. Another possibility is that the X-ray flux from the inner disk is not low because the mass flow rate is low, but because the efficiency of emitting X-rays is low. We will discuss this in the next section.

6. Lessons from the X-Rays

Near the maximum of the outburst, lower energy X-rays from the black hole novae show a component that seems to come from a hot, opaque, geometrically thin disk, as predicted by the disk instability models. The observations show no significant change in the inner radius of the disk as the systems cool after outburst. The only characteristic radius in the disk that could plausibly remain constant as the mass flux declines is the last stable circular orbit, within which matter must plummet straight into the black hole. Evidently, near the peak of the outburst, the accretion disk extends all the way down to the inner radius from which matter plunges directly down to the event horizon of the black hole and disappears. This conclusion strongly affects considerations of the higher-energy X-rays that may contain direct clues of the existence and nature of the black hole, rather than the accretion disk.

The black hole novae also show high-energy X-rays, ranging all the way up to gamma rays. A process known as *Compton scattering* can produce these high-energy X-rays when low energy photons scatter from a hot plasma and pick up energy. Arthur Holly Compton won the Nobel Prize in 1927 for his discovery of this effect and was further honored by the naming of the *Compton Gamma-Ray Observatory* (see Chapter 11, Section 2). Neutron star systems rarely display this kind of radiation, and then only in a truncated form. This high-energy radiation may be just the clue we need to clearly distinguish accreting black holes from accreting neutron stars without the need to invoke the mass limit of neutron stars. Some recent theories for this high-energy radiation have made the explicit argument that it can only exist as it is observed from systems with no hard surface. That argument, if confirmed, would rule out not only neutron stars but also some other bizarre suggestions that would nevertheless have a hard surface. The only small-radius, high-gravity objects we can now imagine that do not have hard surfaces are black holes.

This high-energy radiation is seen near the peak of the outburst of many of the black hole X-ray novae. It probably comes from a hot corona surrounding the disk, although the exact nature of that corona remains elusive. The black hole X-ray novae also commonly show radio outbursts that require an outflow of matter with very high energy electrons. This outflow could also be a source of high-energy radiation. The observed interplay between the high-energy radiation from a corona and the lower-energy X-ray radiation that is presumed to come from the accretion disk is complex and varies in time, but as the outburst decays, the high-energy radiation comes to dominate. This suggests a change in the structure of the accretion flow.

One possibility under active investigation is that, as the mass flow rate declines due to the inward propagation of the cooling wave in the disk, the inner disk thins out and reaches a state where it cannot cool efficiently. Rather than dropping into the cold state of an accretion disk, this inner region can become very hot, and nearly spherical. Matter from this dilute, nearly spherical region then falls almost radially straight down the black hole. The basic notion of this sort of flow was outlined in Chapter 4 and is illustrated in Figure 10.1. This material does not radiate much despite its high temperature both because dilute gas does not radiate efficiently and because this matter tends to plunge directly down the black hole, carrying its heat energy with it. In these circumstances, there is little time to radiate. This process is called *advective accretion flow*, to distinguish it from disk accretion flow. In a disk, most of the heat energy is radiated out through the face of the disk. In an advective flow, the heat is carried, or advected, down through the event horizon, so little heat is lost to radiation.

What little heat does radiate from an advective flow should, according to theoretical models, emerge as very high energy radiation, as observed. Because the radiation efficiency is low, a much higher mass flow rate must be sustained in order to produce even the feeble radiation that is seen. When applied to the black hole X-ray novae, this theory suggests that a substantial amount of the mass transferred

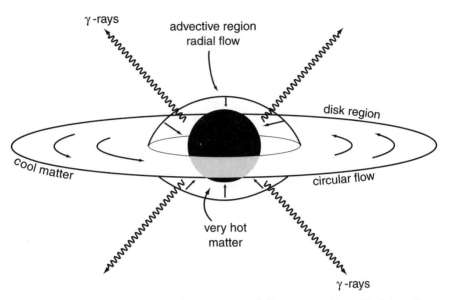

Figure 10.1. To account for the high-energy radiation observed from black hole X-ray novae as they enter the low-luminosity state, a nearly spherical central advective region may form where the flow is nearly radial and the matter is very hot, but radiates inefficiently. Matter from the companion star spirals down through the accretion disk and then, perhaps by a process of evaporation, joins the hot advective flow before plunging down the black hole.

from the companion star does pass through the disk and down the black hole even when the system is in its long-lived, low-luminosity state. Models based on this picture are rather successful in accounting for the feeble X-rays from the low-luminosity systems, even though the simple disk models say the disk should be cool and in a storage phase. This theory is on the cutting edge of research as this book is being written and so there are a number of questions that have not been completely resolved. Among these are: how a cold disk can pass all the mass it must in order to feed the advective flow; how the advective region forms, perhaps by evaporation of disk matter; whether a substantial amount of matter transferred from the companion is blown away in a wind or other outflow before it can reach the black hole. All these issues are a sign of a vibrant and exciting research area.

One general notion has emerged. If the accreting object had a hard surface, photons from that surface would probably interfere with the matter in the advective region and prevent it from having the properties observed for the black hole sources. This is one version of the argument that the black hole X-ray novae cannot be neutron stars but must be objects with no surface. If this argument is right, they must be black holes, independent of the mass we measure for them.

7. SS 433

Another interesting class of objects in the astronomical zoo consisted for a very long time of a single entry. In 1980, Walter Cronkite brought this discovery to the attention of the world when he announced on CBS News that astronomers had found an object that was coming and going simultaneously! For those of you confused by that, read on.

The object was originally identified as being notable for its *emission lines*, excess power coming out at certain wavelengths of light. Normal stars show absorption by cool atoms, and emission is a sign of an energetic environment in some fashion. The object at issue is source number 433 in the catalog of objects with strong emission lines compiled by two astronomers, Sanderson and Sanduleak, so it is known as SS 433 (this is the same Sanduleak who cataloged the star destined to erupt as SN 1987A). Closer study showed that the emission lines in this object displayed a most peculiar behavior. There are two sets of emission lines, and they move around in frequency in opposite directions because of the Doppler effect. Each set of lines shows first a red shift and then a blue shift. The period of oscillation is 64 days. When one set of lines shows a red shift, the other set shows a blue shift, and vice versa. Thus when the gas causing one set of emission lines is moving toward us, the gas causing the other set is moving away from us, hence Cronkite's comment on the news. The actual interpretation that astronomers have given to this information is that SS 433 is emitting jets of material in opposite directions, but somehow twisting around to throw the beams first in one direction, then in the other. Radio observations show an arcing series of blobs extending out beyond the object. Imagine that you are pointing a water hose overhead, but moving the nozzle in a circle. If you were to take a photograph at one instant, you would see blobs of water strung out along a widening helical path. That is what the radio astronomers see, confirming the picture of the oppositely directed rotating jets.

The real excitement came with the deduction of the velocity of the jet material. The jets are not directed at the Earth, but sideways, so normally one would not expect a Doppler shift. According to Einstein's special theory of relativity, however, even an object moving sideways shows a tiny Doppler effect. With ordinary velocities, the effect is undetectable. In order for there to be a measurable "transverse" Doppler effect in SS 433, the material in the twin beams must be moving at 80 percent the speed of light! SS 433 is ejecting opposing beams of material at nearly the speed of light. Active galaxies and quasars had shown similar jets, but this was the first time a star displayed such phenomena.

A further remarkable feature is that the material in the beams is not hot. SS 433 shows emission lines of neutral helium, but none from ionized helium so the matter cannot be tremendously hot. How the matter accelerates to the speed of light without getting heated in the process is a question that still plagues the theorists. One possibility is that radiation pressure can slowly accelerate the material and never push on it so hard that it gets hot.

SS 433 is surrounded by a radio source identified by the *synchrotron radiation* that arises when electrons spiral around in magnetic fields at nearly the speed of light. Some have identified this radio source as a supernova remnant left from the formation of SS 433. Others point out that if this is so, it is the largest supernova remnant in the Galaxy. A plausible alternative is that the remnant is a bubble blown in the interstellar gas by the relativistic particles ejected in the twin beams of SS 433 itself.

The actual nature of SS 433 still eludes satisfactory explanation. Clearly, the tremendous velocities require high energy and thus probably the high gravity of a compact star. One idea is that SS 433 contains a neutron star that is trying powerfully to emit radiation, perhaps because it is a young and energetic radio pulsar. If mass transfer has totally enshrouded it in a blanket of gas, a common envelope, however, the radio waves could not get out directly. The energy then blasts out of two holes in the top and bottom of the envelope and makes the beams. This notion is given some support by other Doppler shift measurements that indicate that besides the rotation of the beams, the whole object moves about with a period of 13.6 days. This probably represents a binary orbital period. The binary companion is presumably the source of the enshrouding envelope. Other theories attribute the energy to matter being swallowed by a black hole.

SS 433 remains an enigma in many regards, and the search for another object like it anywhere in the Universe went on for over a decade. Its close cousins, if not twins, were discovered only a few years ago.

8. Miniquasars

The black hole X-ray novae discussed in Sections 4 and 5 drew a lot of attention as evidence grew that they were black holes. The specifics were different in detail, but these objects had an inflow of matter, accretion disks, and, very probably, black holes. The same general description applies to the models for the energy sources of quasars and active galactic nuclei. The main difference is that the black holes in quasars are thought to be supermassive, up to a billion solar masses, and those in the black hole X-ray novae were 5–10 solar masses. The latter were clearly formed by the collapse of stars (although the details elude us). We do not know the origin of the supermassive variety.

One aspect of the supermassive black holes in galaxies is that they often emit beams of matter at nearly the speed of light. SS 433 was a hint in the direction that stellar mass black holes could do the same thing, but ambiguity about its nature prevented a direct analogy from being drawn. That situation changed dramatically in the mid-1990s with the radio study of the outbursts of some of the black hole X-ray novae.

Felix Mirabel is a radio astronomer of Argentine extraction who works in Paris. Luis Rodriguez is a Mexican radio astronomer. They began a project to monitor the radio emission of the black hole X-ray novae. In 1994, they got data on an outburst in an otherwise obscure source that is hidden behind so much galactic dust that it cannot be seen with optical telescopes. The radio emission can penetrate the dust. Mirabel and Rodriguez discovered a remarkable behavior. They could identify discrete clouds of particles ejected from the X-ray source that emitted radio radiation as they moved rapidly away from the central source. By watching these clouds from day to day, they could see how far apart they had moved in a given time interval. A simple calculation of their speed showed that they seemed to be moving at greater than the speed of light!

This apparently superluminal behavior had been seen before. It was first noticed 30 years ago when similar monitoring was done of quasars. This does not represent a breakdown of Einstein's theory, but a sort of relativistic optical illusion. The explanation for this phenomenon gave Sir Martin Rees, the eminent British astrophysicist, his first claim to scientific fame. The answer to this puzzling behavior is that the matter is ejected from the central source at nearly, but not quite, the speed of light. For the sources that appear superluminal, the jets of matter are pointed nearly toward us. In this case, the matter is chasing the radiation it emits and traveling at nearly the same speed. This foreshortens the apparent motion of a blob of emitting matter in such a way that it seems to be covering a large angle, and hence a large reach of space, in an impossibly short amount of time. The X-ray nova that Mirabel and Rodriguez observed was doing the same thing. The matter was being ejected in blobs that moved at nearly, but not more than, the speed of light, thus giving the appearance of superluminal motion.

At least one other black hole X-ray nova has been discovered to display this superluminal motion. The second one has a measured mass for the compact object from the binary orbit that it is more than 3 solar masses. This puts it firmly in the category of black hole candidate. The miniquasars have helped to put SS 433 in context. There are differences, but there are also obvious similarities. Even though there is still no firm proof that SS 433 is a black hole, we can deduce that if the jets of SS 433 were pointed more nearly directly at us, we would witness nearly, if not clearly, apparent superluminal motion.

The analogy between the black hole X-ray novae and quasars as supermassive accreting black holes was already quite strong, but the discovery of the X-ray novae with apparent superluminal motion cemented the idea in many people's minds. The phrase "mini-quasars" instantly became popular to describe the black hole X-ray transients, especially those with the superluminal behavior. There is much to be learned about how black holes of either the stellar or supermassive variety launch the rapidly moving blobs of radio-emitting matter, but the discovery of the miniquasars is one more piece of evidence that black holes really exist on both the stellar and supermassive scales.

11

Supernovae, Gamma-Ray Bursts, and the Universe

Long, Long Ago and Far, Far Away

1. Probing the Size, Shape, and Fate of the Universe with Supernovae

1.1. Supernovae as Signposts

Apart from their intrinsic interest as star-destroying explosions, supernovae have other uses simply because they are so bright. Their great luminosity means that they are visible across the Universe. More specifically, supernovae are signposts that determine the distances to their host galaxies. Careful measurements of those distances allow astronomers to map out how fast the Universe is expanding and hence how old it is, the curvature of space, and clues to the fate of the Universe. The use of supernovae in this way has expanded extensively in the last few years and the results have been dramatic. Supernovae have provided clues that the Universe may expand forever and that it is even now in the grip of powerful repulsive forces that accelerate its outward rush.

To use supernovae as a technique to measure distances requires some perspective on what we are trying to accomplish and how we are doing the task. Recall from Chapter 9 the various two-dimensional analogs we have employed to picture curved space. The two-dimensional space around a gravitating object is funnel-like when viewed from the perspective of three dimensions. The two-dimensional analog of the Universe itself, at one moment of time, can be represented as the surface of a sphere, an infinite flat plane, or a saddle extending upward to infinity fore and aft and downward to infinity sideways, as shown in Figure 11.1. These two-dimensional analogs are the embedding diagrams for the Universe. They help picture curvature in three dimensions. We have stressed that looking down from a higher, three-dimensional perspective is cheating in a sense because there is no way we can look down on our three-dimensional curved space from an "outside." That outside would, by analogy, itself have to be a fourth spatial dimension. If there were an observer in that fourth spatial dimension, that observer could see the curvature of our Universe or that around the Earth or around a black hole in much the same way that we can see the curvature of the surface of a sphere. On a more direct and personal level, such an observer would also not be limited to viewing our surfaces, our skin, and our facial features as we do one another. An observer from a hypothetical fourth dimension would also be able simultaneously to see our volume, our guts, and our bones, much as we can see the interior of a circle inscribed on a sheet of paper. This is an amusing perspective, but it is not one of physics.

Rather, the proper perspective is to recognize that a two-dimensional creature living in any of these curved two-dimensional spaces could determine that the space curves and by how much by doing geometry, by carefully measuring distances and angles. That is now our task! We are three-dimensional supernova observers trapped in our three-dimensional Universe. We must determine the curvature of our three-dimensional space without stepping outside of three dimensions, something we simply cannot do. Fortunately, we do not need to step outside. We just have to be careful with our geometry and our astrophysics.

1.2. Type Ia Supernovae as Calibrated Candles and Understood Candles

The use of supernovae to measure distances is based on a simple principle: things farther away look dimmer. Turned around, how dim a supernova appears to be is a measure of how far away it is. The basis for this intuitively reasonable notion is that, when light spreads out from a central source equally in all directions, the locus of the photons emitted at a given time defines a larger and larger surface. The light falling on a detector of a given area, a human eyeball or a

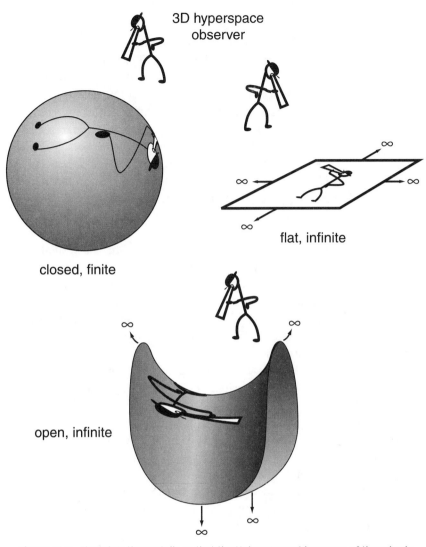

3D hyperspace observer

closed, finite

flat, infinite

open, infinite

Figure 11.1. Einstein's theory tells us that the Universe must have one of three basic shapes. The two-dimensional analogs (embedding diagrams) for these cases are a spherical surface (a "closed universe"), a flat plane extending to infinity in all directions (a flat universe), and a saddle shape that also extends to infinity in all directions (an open universe). Two-dimensional astronomers in two-dimensional universes cannot stand outside their universes to see the nature of the curvature the way a three-dimensional hyperspace observer can. Rather, they can do geometry in the context of their own space and determine the shape of their universe. Triangles in flat space will have their interior angles sum to 180 degrees, but the answer will be more than 180 degrees in the spherical universe and less than 180 degrees in the saddle-shaped case. As three-dimensional astronomers in our own three-dimensional Universe, we cannot stand outside of it in hyperspace, but we can do geometry to determine the nature of the Universe we occupy.

telescope, then captures a smaller and smaller fraction of the total the farther away the detector is from the source. The fraction decreases just as the total area into which the radiation floods increases and that goes like the distance squared (the area is $4\pi D^2$, where D is the distance). This means that the apparent brightness of a source of a given total luminosity decreases like the inverse of the square of the distance. In simple terms, the fainter a given kind of object appears, whether it is a porch light, a star, or a supernova, the farther away it must be. If you know how bright the object really is, then you can tell from how bright it apparently is how far away it must be. This gives us a powerful tool for measuring distances. The key is to figure out how bright a given object really is.

Recall that Type Ia supernovae are generally the brightest of all the different types. This makes them especially good signposts for measuring large distances. If we knew exactly how bright they were, the task of measuring distances would be rather easy. We would just see how bright a supernova looked in a given telescope and read off the distance. The immediate problem is to determine the intrinsic brightness of a given supernova.

For a long time, there was some reason to believe that Type Ia supernovae were all equally bright. That would have made the task of measuring their distances particularly easy. The jargon for this is that such identical supernovae would represent a *standard candle*. The idea is that, if you have a set of "candles" of identical, known brightness, they can serve as a "standard" with which to compare other sources of luminosity and to measure distances. In the last decade, we have determined that Type Ia supernovae are not exactly the same, but that the differences are systematic. That allows astronomers to make allowances for the differences between individual Type Ia supernovae.

In particular, astronomers have found that the Type Ia supernovae that are intrinsically brighter decline in brightness more slowly than those that are intrinsically dimmer. We believe that we even have a basic understanding of why this is true. Some variation in the exploding white dwarf causes variation in the amount of radioactive nickel-56 produced in the explosion. The extra energy from radioactive decay does not just make the supernova brighter, it also keeps the expanding matter opaque longer. The radiation takes longer to leak out, giving the slower decay. The trend that relates the brightness of the supernova to the rate of decline from peak light gives the means to determine the brightness of the supernova. One just needs to see how fast the supernova declines, and that tells you how bright it really is. Comparison with how bright it seems in the telescope then gives the distance.

There are two ways of doing this comparison. One uses only the empirical data from the supernova with no attempt at a theoretical understanding. This method requires some comparison with other astronomical objects for which the distances are established in some other way. This calibration sets the overall scale of just how bright a Type Ia supernova with a given rate of decline really is. This must be done for as many supernovae as possible for which the distance is already known (in practice a dozen or so, with the sample growing as this book is being completed). Then the brightness-decline relationship gives the intrinsic brightness and hence the distance from a measurement of the decline rate alone. This technique uses Type Ia not as standard candles but as light sources for which the brightness of each supernova can be calibrated compared with known sources, hence the phrase "calibrated candles."

The other technique to employ Type Ia supernovae to measure distances uses theoretical models of the explosions to determine how bright the supernova must be to produce a given light curve. This technique thus attempts to employ "understanding" rather than "calibration" to provide the necessary information to turn the decline rate into a known intrinsic brightness. This technique thus uses Type Ia supernovae as "understood candles."

The first technique, using the Type Ia supernovae as calibrated candles, is only as good as the calibration and the implicit assumptions that underlie the empirical relation between peak brightness and the rate of decline. A key assumption is that the brightness-decline relation is unique. Two supernovae with identical decline rates are assumed to have the same intrinsic peak brightness. The second technique, using Type Ia as understood candles, is only as good as the rather complex underlying theory of the explosion and of the production of luminosity. This method can, in principle, allow for cases where, because of more subtle circumstances, other variables enter and two supernovae with the same decline rate do not have the same peak brightness. The two methods agree rather well. They both give the same age of the Universe.

1.3. The Age of the Universe

The Universe we see around us began in what we call the big bang. There are still mysteries surrounding how the Universe came to be. We will touch on some of them in Chapter 12. There is, however, no doubt that the visible Universe arose in a very dense, hot state, and expanded outward. Although the first instants are murky, ordinary particles, protons and electrons formed very quickly, and the Universe was pure hydrogen for a while. The light elements – helium, lithium – formed when this expansion was a few minutes old. When it was a million years old, the matter got sufficiently dilute that the radiation from its heat could stream freely. We see that radiation as the *cosmic background radiation* that comes at us from all directions. This cosmic radiation is red-shifted by the expansion that pulls everything in the Universe away from everything else. We understand this process very well. Further expansion of the Universe brought the agglomeration of matter into galaxies, stars, and planets in ways we are still striving to understand. Continued expansion pulls all the distant galaxies apart. Understanding the expansion of the Universe allows us to measure its age.

It is important to realize that the big bang did not occur as an explosion in a preexisting space, like a bomb in outer space. Rather space itself expanded, carrying the matter with it. One popular analogy is the behavior of spots on the surface of an expanding balloon. The spots do not move with respect to the rubber surface as the balloon expands, but they become ever farther apart, as shown in Figure 11.2. A three-dimensional analogy is raisins in a rising loaf of bread. The raisins never drift in the dough but again move ever farther apart until the loaf stops rising. The second analogy is limited and a little deceptive because the loaf of bread is finite. The three-dimensional loaf of bread is surrounded by ordinary three-dimensional space into which it expands, whereas the space of the Universe is all-encompassing. The first analogy is limited because it is restricted to two dimensions, but it is more revealing in a way. One can see that the two-dimensional surface of the balloon has no two-dimensional outside, neither the outside as we understand it from our three-dimensional perspective nor what we regard as inside the balloon, which still requires going off into a third-dimensional "hyperspace" from the perspective of a two-dimensional creature inhabiting the two-dimensional surface. Likewise, the loaf of bread is perceived

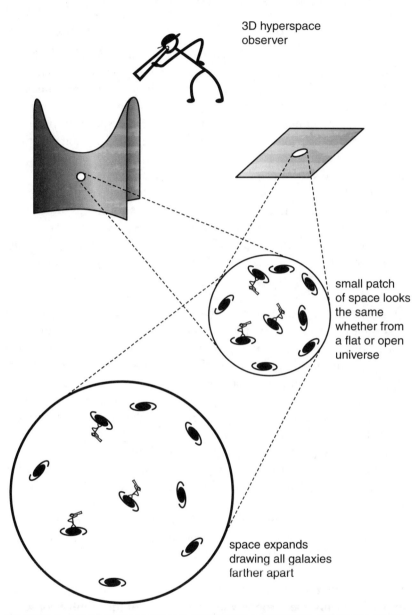

3D hyperspace observer

small patch
of space looks
the same
whether from
a flat or open
universe

space expands
drawing all galaxies
farther apart

Figure 11.2. A small piece of any two-dimensional universe will appear flat. As the universe expands after its big bang, this piece of the universe will expand, drawing all the galaxies in it farther apart with time. A three-dimensional hyperspace observer could see this expansion, but two-dimensional astronomers resident in the two-dimensional universe could determine the expansion by registering the Doppler red shift as all distant galaxies move apart from all others. As three-dimensional astronomers in our own three-dimensional Universe we cannot stand outside, but we can measure Doppler shifts of distant galaxies and determine how fast the Universe is expanding.

to have a center, whereas (ignoring the opening though which one blows) there is no two-dimensional center to the two-dimensional surface of a perfect sphere to which the balloon is an approximation. Unlike the loaf of bread, the balloon shows that if attention is restricted to the confines of the dimensions of the space, two for the surface of the balloon, three for our Universe as we perceive it, there is no center and there is no outside. These are tricky and fascinating issues, and we will return to them in Chapter 12.

For our current purposes, it is sufficient to picture the expansion of the balloon and its dots or the bread and its raisins to understand how to measure the age of the Universe. The effect of the expansion of the Universe is still much the same as an explosion in preexisting space even if the concepts are radically different. If you can measure how far away something is from you, say a distant supernova, and determine how fast it is traveling away from you, by measuring its Doppler shift to the red, then you can tell how long it has been traveling to get as far as it has. You get the same answer for every supernova and every galaxy. The faster they move away from us, the more distant they are, but they took the same time to get there, drawn by the expansion of the underlying space.

The parameter that is measured in this way is called the *Hubble constant*, after Edwin Hubble who pioneered this sort of measurement of distances and determined the nature of the Universal expansion. The Hubble constant tells you how fast something will be moving away from you at a given distance. Both techniques for measuring the distances to Type Ia supernovae outlined in Section 1.2, and other techniques as well, say that velocity will be about 65 kilometers per second for every million parsecs in distance. The age is related to the inverse of the Hubble constant. Obtaining the age of the Universe from the Hubble constant involves another subtlety because it depends on the curvature of space and the acceleration of the Universe. Neglecting that subtlety for the moment, the corresponding age of the expanding Universe is roughly just the inverse of the Hubble constant. If a supernova moving at 65 kilometers per second is 1 million parsecs away, it must have been moving away from us for about 10 billion to 15 billion years. If another supernova is moving away from us at 650 kilometers per second and is at 10 million parsecs, then the time for it to get there is just the same, 10 billion to 15 billion years. We get the same answer for every supernova, as we must because we are measuring the same age in every case, the age of the Universe.

1.4. The Fate of the Universe

The game is not over with the measurement of the Hubble constant. It is not enough to measure how old the Universe is. We want to know what will happen to it in the future. Since the days of Hubble, astronomers, particularly the subset known as cosmologist, have been engaged in a grand quest to determine the "fundamental parameters of the Universe." This quest was shaped by Einstein's theory of gravity. The first attempts to apply Einstein's theory to the whole Universe showed that there were three parameters that would describe the whole shebang: the Hubble constant, the overall curvature of the Universe, and the rate at which the Universe is changing its speed of expansion due to the gravitational pull of the matter and energy within it. The issue of curvature is whether the Universe is the three-dimensional analog of the surface of a sphere, a flat plane, or a saddle, as shown in Figures 11.1 and 11.2. Einstein's theory showed that it had to be one of the three. Furthermore, with a key simplifying assumption, the theory showed that if the Universe were spherelike, it would have a finite life and recontract to a singularity; if it were flat, it would expand forever, just reaching zero expansion rate at the end of time; and if it were saddle-like, it would expand forever at a finite velocity. The models of the Universe that defined these fundamental, and measurable, parameters made some simplifying assumptions. We will see later in this chapter and in Chapter 12 that these parameters may not tell the whole story, but they make up a critical part of it. Determining these parameters occupied cosmology for most of the twentieth century.

There are various ways of going about measuring the other two parameters in addition to the Hubble constant. Using supernovae, the underlying theory requires the constraint of two specific quantities. One is the mass density of the Universe at the current epoch. In its simplest guise, this means determining the total mass of all kinds of stuff that has a finite mass and does not move at the speed of light. This mass includes stars, planets, and dust, but it also means any component of the mysterious *dark matter* that consists of particles, no matter how exotic. The other quantity to be constrained (and ultimately measured) is the value of what is called the *vacuum density*. Recall that even a vacuum has an energy associated with it. This energy underlies the emission of Hawking radiation from black holes. The vacuum may have even more subtle properties that would only be manifested when its effects are determined on the scale of the whole Universe.

There is a story behind this vacuum energy. The vacuum energy is, in principle, related to the quantum properties of the vacuum, but it arises in Einstein's theory of gravity where it is called the *cosmological constant*. Astronomers who write the history of this subject tend to quote Einstein himself in this regard with great glee. Einstein called the cosmological constant "the greatest blunder of my life." The historian's glee and Einstein's self-criticism are probably unfair. The cosmological constant emerges from the mathematics of Einstein in a perfectly natural way (it appears as a constant of integration, for those who know calculus). It is not a question of whether it exists in this mathematical sense. It certainly does. The issue is whether it is zero or not, and whatever its value, including zero, what the physics is that determines that value.

The reason Einstein regarded his treatment of the cosmological constant to be a "blunder" is that the first mathematical models for the Universe showed that the Universe would expand. Einstein's intuition told him that the Universe could not possibly do such a radical thing. To render the solution static, Einstein went back to the equations and realized that he had implicitly set the value of the cosmological constant to zero. If he assigned it just the right nonzero value, then the cosmological constant could serve as an extra source of gravity and balance the tendency of the Universe to expand. Shortly afterward, Hubble proved that the Universe *is* expanding. It appeared to Einstein that the cosmological constant was unnecessary, a blunder.

Einstein may have blundered in guessing that the Universe was static, and hence in the value to which he set the cosmological constant, but he did not blunder in introducing the idea. In the long run, it is the latter that is more important, and another tribute to the power of Einstein's theory. The blunder was much less than it is often made out to be. We now see that even the issue of whether the cosmological constant might be exactly zero is not a trivial one, but one that involves some of the deepest thinking about the Universe. More than that, there are hints that the cosmological constant is not zero, and that definitely raises profound issues of physics and cosmology.

1.5. Supernovae and Cosmology

Using supernovae to determine the other fundamental parameters of the Universe has been a dream for decades. Many people have worked for a long time to bring it to pass. One of the pioneers, Stirling Colgate of the Los Alamos National Laboratory, estimated that to get the job done when he started working on an automated supernova search telescope in the early 1970s, he would have had to invent seven or eight brand new technologies. These included digital control of the telescope and its instrumentation, electronic detectors to replace photographic plates (Colgate called all this "dig-as" for digital astronomy; the tide of the digital revolution has fully enveloped astronomy by now, but the term never caught on), thin lightweight mirrors, time-sharing computers necessary for many people to work cooperatively on the complex computer code required to control the telescope and scan images, and cheap microwave links to allow remote control of the telescope from a distant site. The telephone company wanted $3 million for a microwave link from his telescope to the headquarters in Socorro, New Mexico. Colgate had only $3,000 for the job. He invented a simple method of error checking and installed the link with the funds and equipment he had.

In just the last few years, the technical capability, the development of critical techniques, and the willingness to devote a great deal of hard work have come together to bring this dream to fruition if not in quite the fully automated way Stirling Colgate envisioned. A key development has been the construction of large new telescopes and the special electronic detectors to record faint images over relatively large patches of the sky. Another was the launch, repair, and updating of the *Hubble Space Telescope.*

A team of astronomers at the Lawrence Berkeley Lab of the University of California, now headed by Saul Perlmutter, pioneered the breakthrough in technique. One of the inhibitions of research on supernovae is that their eruption is always a surprise. This means astronomers have to scramble to get data when an explosion occurs. Telescopes are often in the wrong configuration with the wrong instrumentation, the Moon is too bright to see the faint supernova light, or the weather is poor. The result is that we still do not get adequate information on most supernovae.

The Berkeley team realized that in certain circumstances they could discover supernovae "on cue." They could then schedule procedures in advance to follow them up. These techniques work in precisely the context where one can use the resulting discoveries to do cosmology with supernovae. The trick is that if one looks out to very large distances, a given image obtained with a telescope spans a huge volume containing a huge number of galaxies. It is impossible to predict which of the many galaxies will produce a supernova, but if enough galaxies are in the image, one can be confident that some supernovae will erupt. It turns out that one does not even have to know which specific galaxies are there in advance. If one looks distant enough, there will always be plenty of galaxies and plenty of supernovae. The distances involved, billions of light years, are also just the distances astronomers needed to probe to learn about cosmology.

More particularly, the technique developed by the Berkeley team is to schedule time on a large telescope when the Moon is not up and the sky is dark. They obtain a first image of the sky. They then return and take another image of the same patch of sky 2 or 3 weeks later after the Moon has passed through its bright phase and is no longer a problem. They compare the second image to the first and look for any new lights in the faint images. This is not trivial because both the galaxies and the supernovae are very faint. Many person-decades have been invested in the computer codes that can automate this process and detect and eliminate flashes of man-made light, cosmic rays that strike the detector, asteroids, and other things that are just a nuisance for this project.

Nothing can be done about bum weather, but these procedures have brought the other factors under control. In addition to the Berkeley group, another group has sprung up in competition led by Brian Schmidt of Mt. Stromlo Observatory near Canberra and comprising astronomers in Chile, at Harvard, and elsewhere. The results have been striking. The two groups of astronomers can guarantee the discovery of roughly a dozen very distant supernovae each time they return to take the second image. Because they know far in advance when they will take the second image, they can coordinate the prior scheduling of other telescopes. In this way, they are prepared to get critical spectral and photometric information as soon as they determine the precise location of the new discoveries. Rapid global communication, including the Internet, also plays a key role here. Both teams have also used the *Hubble Space Telescope* to closely examine the host galaxies after the supernovae have faded. This is a critical step because one must subtract off the light of the host galaxy to get a pure signal from the supernova. Determining the light of the galaxy alone can be done efficiently after the supernova has faded, but not when the supernova first goes off and the light is a complex admixture of supernova and galaxy emission. This technique requires patience. Several months must pass before the supernova has faded sufficiently, and many more months are required for careful calibration and analysis. Using these techniques, the number of supernovae discovered per year has shot up to around 100, most of them at distances that span a good fraction of the observable Universe.

The results of these efforts are just now becoming known. Taken at face value, they are very surprising. Recall from Section 1.2 that for a given intrinsic luminosity, the apparent brightness of a supernova declines as the inverse of the distance squared. This result, like the ratio of the circumference to the radius of a circle and the sum of the interior angles of a triangle, depends on the curvature of the underlying space. The power of the method of using supernovae is that they can, in principle, give such precise measurements of the distance at such great distances that the effects of the curvature of the space can be gleaned.

As mentioned earlier, the amount of mass of all kinds in the Universe affects the curvature of the Universe and tends to slow down the expansion because of the mutual self-gravity of all the mass-energy. If the Universe is slowing down, then it was expanding more rapidly in the past. This means that, when we look at supernovae long, long ago and far, far away, with a given Doppler red shift, they will be a little closer and a little brighter than if the Universe had just been coasting at a constant speed, as shown in Figure 11.3. The Universe will also be younger than one would estimate from a given value of the Hubble constant and the assumption that the Universe had always expanded at the current rate.

That is all there is to it if the value of the cosmological constant is zero. If the cosmological constant is not zero, then the effect depends on whether the cosmological constant is positive or negative. If it were negative, the energy of the vacuum would add to the gravity of the matter and slow the expansion even more. If the cosmological constant were positive, the vacuum energy has the effect of a repulsive force causing the Universe to fly apart ever faster as it ages. That sounds like a strange effect, but it is possible within the framework of Einstein's theory and another measure of why the introduction of the cosmological constant was not a blunder but a very fascinating step.

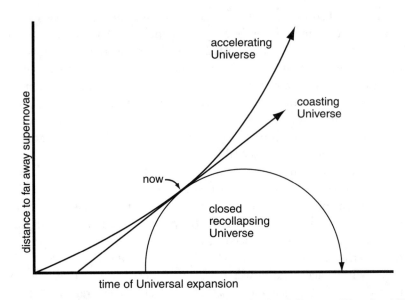

Figure 11.3. The size of the Universe as measured by the distance and Doppler shift of distant supernovae as a function of the age of the Universe. The three lines represent, schematically, the behavior of a closed Universe that is destined to recollapse, a flat Universe that will slowly coast to a halt in infinite time, and an accelerating Universe. The lines all have the same slope at the epoch marked "now." The slope of the lines at that point gives the Hubble constant. The beginning of the lines represent the origin of the big bang for each case. For a given slope of the lines now, the closed Universe gives the shortest time since the big bang, and the accelerating Universe gives the longest.

The mass density in the Universe must be positive, but the value of the cosmological constant could be positive or negative or zero and must be determined by observation or theory. If the cosmological constant were positive, it would act in the opposite way to the mass density. A positive cosmological constant would tend to make the Universe accelerate rather than decelerate, as shown in Figure 11.3. This means that a supernova at a given red shift will be a little farther away and a little dimmer than if the Universe had expanded at a constant rate. Likewise, the Universe would be a little older than one would estimate for a given Hubble constant and the assumption of a constant rate of expansion.

Because the effect of the positive mass density and of a positive cosmological constant work in opposite directions to determine the dynamics of the Universe, the measurement of distances to supernovae tends to constrain the difference between the two effects. Using supernovae alone, the effects cannot be easily separated. Careful measurement of the apparent brightness and red shift of Type Ia supernovae of a given rate of decline and hence intrinsic brightness can, however, constrain the values of the mass density and the cosmological constant. From those constraints and a knowledge of the Hubble constant, the curvature of space and the rate of change of the speed of expansion of the Universe can also be estimated.

The current work on distant supernovae has given two surprises. One is that there does not seem to be enough matter to close the Universe. This suggests that the Universe will expand forever. There are other astronomical techniques that are giving the same result. They all need to be further refined and considered, but astronomers are taking the possibility seriously. There is an even more surprising result from the supernova work. Compared to the local sample of supernovae on which the calibration is done and compared to a Universe for which the cosmological constant is zero, the distant supernovae seem to be a bit too dim. If this is caused purely by cosmological effects, then the implication is that the supernovae are a bit farther away for a given red shift. This effect, in turn, can only be explained by a finite and positive cosmological constant or an equivalent effect. The implication is not only that the Universe is not closed and finite but also that it is accelerating, not decelerating! This is a striking and unexpected result.

A positive cosmological constant raises profound questions about what the nature of the vacuum must be that it contains a quantum property that acts as a repulsive force. The most popular models of the big bang now are so-called "inflationary" models. In this picture, when the Universe was first born, it had a vacuum energy that did act as a repulsive force, an antigravity, that caused a piece of the Universe to expand rapidly to form the Universe we see today. According to the theory, this energy of the vacuum should have decayed away to zero by now. If the vacuum still has some of this repulsive energy, new theories of the vacuum will have to be developed.

This discovery has also upset the cosmological game plan to discover the fate of the Universe by measuring the three fundamental parameters of cosmology as described in Section 1.4. It remains true that determining, directly or indirectly, the Hubble constant, the matter density, and the cosmological constant, one can determine the shape of the Universe, open, closed, or flat. If there is a cosmological constant, however, that information alone may not reveal the fate of the Universe. If the Universe has a low density as current results suggest and a positive cosmological constant so that the tendency to coast outward is even accelerated, then infinite expansion is certainly suggested. In principle, however, a positive cosmological constant could continue to push the Universe into infinite expansion even if there were enough matter to close it, which there does not appear to be. On the other hand, given that the existence of a cosmological constant raises issues of its origin that we clearly do not know how to answer, we cannot be sure that the cosmological constant is "constant." If this vacuum energy should switch signs and the cosmological constant become negative, then, again in principle, the Universe could be doomed to recollapse even though it did not contain enough matter to accomplish that feat on its own. These results have opened up new, if misty, vistas in both cosmology and physics.

These supernova-based results are still new as this book is being written and need to be carefully considered. One of the critical issues is whether it is not the properties of the Universe that were different in the past, but the properties of the supernovae themselves. We know that not all Type Ia supernovae are alike and that their properties tend to correlate with the age of the stellar systems and host galaxies from which they emerge. The teams analyzing the supernovae have made some allowances for this. The question is whether the allowances are adequate. The current techniques to determine the brightness of the distant supernovae use the calibrated candle methods. An important task will be to check this result with the understood candle method. In addition, both of these methods currently use only rather gross properties of the supernovae, how much light is emitted at a given time, to determine the properties of the event. Much more information about subtle differences is contained in the details of the evolution of the spectrum of each event. This information has not yet been effectively tapped, but will be in the next few years. These early results of the distant supernova searchers are very stimulating. The implied results of low mass density and a finite, positive cosmological constant may even be correct, but it will probably take several years of hard work by many people before we know that with confidence.

Other techniques will also come into play in the future to determine the state of the Universe. The *Cosmic Background Explorer* (*COBE*) satellite launched in 1989 revealed that the background radiation is of an exceedingly well-defined temperature, as expected. *COBE* also revealed faint irregularities in the intensity of the radiation from different parts of the sky. These regularities were also expected and even inevitable, given our understanding of the big bang. The big bang grew out of a singularity. That singularity must have been subject to quantum fluctuations in its properties that would be imposed on the expansion of the Universe and hence on the density and temperature of the matter in the Universe. Detection of these irregularities was another major vindication of the big bang picture.

New satellites to measure the properties of the cosmic background radiation will be launched early in the next century. Careful measurement of the fluctuations in the background radiation has great promise to further constrain the matter density and the cosmological constant. These techniques will be a critical complement to the research based on supernovae. For this effect, the mass density and the cosmological constant tend to work in concert. The larger the mass density, the stronger the gravity and the faster the fluctuations will tend to grow. On the other hand, if there is a finite and positive cosmological constant so that the Universe tends to accelerate, then the Universe will be a little older than it otherwise would be, other things being the same, and this gives the fluctuations more time to grow, again making them larger. The result is that the measurement of the fluctuations in the cosmic background will tend to measure the sum of the mass density and the effect of the cosmological constant, whereas the supernova technique measures the difference between these quantities. Neither technique by itself is apt to give the full picture. If, however, we have independent measures of both the sum and the difference of the mass density and the cosmological constant, then, in an algebraic sense, we can solve for both unknowns. At this writing, the exercise of joining the data from the supernovae with the data from the background radiation is consistent with the Universe being exactly flat, as the inflation theories demand, with about one-third of the total density being matter (mostly dark matter), and two-thirds being due to the vacuum energy associated with the cosmological constant. Given the remaining uncertainties, however, we cannot rule out that the Universe is barely open or barely closed. Its fate, as mentioned earlier, is even more uncertain.

1.6. The Past in Our Future: The Dark Ages

Looking to the future brings yet another exciting possibility. After the epoch when the Universe was a million years old, the cosmic radiation streamed freely. The matter cooled and became dark. During the subsequent eons of expansion, the matter agglomerated into lumps that became galaxies. At some point, the gas in the lumps condensed and heated and started the first production of stars. The long interval between the release of the cosmic background radiation and the lighting up of the first stars has come to be called the "Dark Ages." After a long period with no light, stars winked on and the Universe started to take the form we recognize around us now. The processes involved in forming the first stars and galaxies, the emergence from the Dark Ages, is one of the frontiers of modern astronomy. It can be probed by the new generation of telescopes in the 8- to 10-meter class. The end of the Dark Ages will be the prime target of the *Next Generation Space Telescope* currently under design by NASA.

Some of those first stars to form will be massive. They will evolve, collapse, and explode in just the way described in Chapter 6. When they do, their host galaxies will still be embryonic, small, and dim. There is a chance that, when astronomers peer back to the beginning of the end of the Dark Ages, they will see supernovae, the brightest beacons in the young Universe.

The first supernovae to arise should be from massive, short-lived stars. They should be predominantly Type II supernovae, although there could also be an admixture of Type Ib and Type Ic supernovae. The Type II supernovae might all resemble SN 1987A by exploding as blue supergiants. As explained in Chapter 7, we do not fully understand why SN 1987A was a blue rather than a red supergiant when it exploded. Theoretical studies have shown, however, that when the amount of heavy elements in the atmosphere of an evolving massive star is low, the hydrogen envelope is likely to remain relatively compact so the star will look hot and blue rather than expanding so that the star will look cool and red. In the very young Universe at the end of the Dark Ages, there will not have been much time to make heavy elements. Whatever caused SN 1987A to be a blue supergiant, the paucity of heavy elements in the young Universe may cause all the exploding stars to be blue supergiants, even if they retain their hydrogen envelopes against the ravages of winds and binary companions.

If the first supernovae at the end of the Dark Ages explode in blue supergiants, the resulting explosions, like SN 1987A, may be relatively dim and somewhat harder to see. As the Universe ages and more heavy elements collect in the interstellar gas from which new stars are born, then at some point, massive stars may begin to evolve into fully formed red supergiants before they die. They will then explode as what we consider to be "normal" Type II supernovae. With the full power of new telescopes to scan from the present epoch back to the end of the Dark Ages, we should be able to see that epoch when the normal Type II supernovae turn on.

This discussion has omitted Type Ia supernovae. That is because we think they have a "fuse" that must burn before they explode. As discussed in Chapter 6, we do not understand the binary evolution that leads to the explosion of a white dwarf as a Type Ia supernova. All the indications are, however, that considerable time must pass before these binary processes, perhaps the evolution of the smaller-mass companion, perhaps the decay of orbits through emission of gravitational radiation, lead to the explosion. That Type Ia supernovae have a long fuse compared to Type II means that when supernovae begin to explode at the end of the Dark Ages, they should all be due to the collapse of the cores of massive stars. There should be no thermonuclear explosions of white dwarfs and hence no Type Ia.

As the Universe ages and the binary evolution fuse burns, there will eventually be an epoch when the Type Ia supernovae begin to explode. Using the big new telescopes on the Earth and in space as time machines to probe these distant times, we should also be able to see this onset of Type Ia events. This would be a very exciting result because the time of the onset will give us critical new information on just what type of binary evolution constitutes the fuse. This, in turn, may finally teach us what binary evolution leads to Type Ia.

2. Gamma-Ray Bursts

2.1. Yet Another Cosmic Mystery

There was a revolution in astronomy in the first few months of 1997. A major breakthrough occurred in one of the outstanding mysteries of modern astrophysics, the cosmic *gamma-ray bursts*. This story began in the 1960s. The United States launched a series of satellites that orbited the Earth at great distance, halfway to the Moon. They were called the *Vela* series, and they were designed to detect gamma rays and other high-energy photons and particles. If it strikes you that there must be something special about them to be so far from Earth, you are on the right track. They were not designed for astronomy, but primarily to detect terrestrial nuclear bomb tests. They were also intended to study the background, other natural sources of high-energy photons and particles in the solar wind and the Earth's magnetosphere, to aid in the separation of bomb signals from natural signals. Stirling Colgate was on the team in Geneva working on the treaty to ban atmospheric tests. He had done some calculations that suggested that when a supernova shock wave broke through the surface of the star there could be a pulse of gamma rays (see Section 2.4 for an update of this topic). He was afraid that such an event would be misunderstood as a nuclear bomb and might trigger a serious miscalculation by one side or the other. He hassled both sides, the United States and the Soviets, concerning the need to understand potential astronomical sources of confusion, lest they lead to disaster. The *Vela* satellites were motivated, at least in part, by these concerns. Colgate found the Russians intractable. They would not do their own astrophysical background checks and feared satellites launched by the United States would be used for spying. The agreement to put the *Vela* satellites in high orbit was a response to the Russian demand for a guarantee that they not be used for spying. Both sides did launch spy satellites, of course, but this did not apply to the *Vela* series, the results of which were unclassified.

Perhaps the *Vela* series saw bombs, but they certainly detected outbursts of an extraterrestrial nature. One of the *Vela* series was instrumented to see X-rays and discovered the first X-ray burst (Chapter 8, Section 7). With the first extraterrestrial detections of gamma rays in 1967 (the *Vela* 4 series), the scientists at Los Alamos could not convincingly rule out the Sun as the source. They had to wait until the launch of the next series (*Vela* 5) in 1969 before they were able to conclude rigorously that the gamma-ray signals were from neither the Earth nor the Sun but from elsewhere in outer space. The discovery was finally announced by Ray Klebesadel, Ian Strong, and Roy Olson in a paper in the *Astrophysical Journal* in 1973. This paper created a new scientific industry.

The bursts of gamma rays from beyond the Earth were seen at irregular intervals. These bursts lasted for 10–30 seconds and showed variations on times as short as a 0.001 second. Subsequent investigations showed that the gamma-ray bursts were primarily a gamma-ray phenomenon, with relatively little energy in the X-ray band, unlike other sources of gamma rays that emit abundantly at lower energies as well. That the dominant emission mode is gamma rays means that a high energy is involved. Gamma-ray bursts probably require high gravity and motion at nearly the speed of light.

The quest for an explanation of gamma-ray bursts was long handicapped by a lack of direct knowledge of the distance to the bursts. A debate raged as to whether they are in the Galaxy or at the farthest reaches of the Universe. This debate was brought into sharp focus by the immensely successful Burst And Transient Source Experiment (BATSE) on the *Compton Gamma-Ray Observatory*. The *Compton Gamma-Ray Observatory*, named for Arthur Holly Compton (Chapter 10), was launched in 1991 as one of the series of Great Observatories planned by NASA. The *Hubble Observatory* was the first. Two others, the *Advanced X-ray Astronomy Facility (AXAF)* and the *Space Infrared Telescope Facility (SIRTF)*, have been downsized, descoped, and delayed for over a decade. *AXAF* has now been launched as the successful *Chandra Observatory*, but *SIRTF* is under new budgetary threats at this writing. BATSE has had many roles, but a principle component of its design was to search for gamma-ray bursts. BATSE recorded over 1,600 new gamma-ray bursts in its latest catalog, which contains data through 1996, and it is still going strong.

The surprising result was that the sources are, to great accuracy, distributed uniformly on the sky. There is no statistical evidence for any tendency to lie toward the plane of the disk of our Galaxy or toward the galactic center. This contradicts any model in which the sources are distributed throughout the Galaxy and viewed from the offset position of the Earth, 25,000 light years from the galactic center. This result fueled increasing conviction that the sources of the gamma-ray bursts were in galaxies at cosmological distances because the distant galaxies are naturally distributed uniformly on the sky, on average. In addition, fainter sources are more abundant. The precise number of faint sources shows a pattern that is close to what one would expect if the bursts constituted a gamma-ray standard candle viewed in ever-larger volumes of space in an expanding Universe. There might, however, be other explanations for this pattern. There is no particular reason to think that gamma-ray bursts are a standard gamma-ray candle.

The problem is that if the gamma-ray bursts are at cosmological distances, the intrinsic source of energy must be huge, comparable to or exceeding that of a supernova, but radiated essentially entirely in gamma rays. The energy requirements are more moderate if the energy is directed in a beam, but then the number of sources must be correspondingly higher. Although some models of colliding neutron stars or neutron stars colliding with black holes or black holes consuming magnetized accretion disks produce such energies, such models tend to strain credibility. Everything about the cosmic gamma-ray bursts strains credibility, yet there they are.

One of the clearly defined problems in the study of gamma-ray bursts was the complete lack of counterpart events at other wavelengths, especially optical wavelengths. Without optical counterparts, the full weight of astronomical lore, much of it derived from optical astronomy, could not be brought to bear on the issue. The problem was that the gamma-ray detectors could not provide sufficiently good locations. It is a difficult technical feat to bring gamma rays to focus. The gamma-ray sky has typically been "fuzzy," a situation somewhat analogous to nearsighted people looking around with their glasses off. A given gamma-ray burst could be said to be "over there," but "there" could not be precisely defined. The uncertainties in position are typically several to tens of degrees in radius (the full Moon subtends about 0.5 degree in angular diameter). In an area of the sky of that size, there can be thousands of stars. Finding the point of light that corresponds to a given 10-second-long gamma-ray burst has been like seeking the proverbial needle in a haystack, a needle that was likely to vanish if you did not find it in less than a minute.

The nature of these events has puzzled astrophysicists for nearly 30 years. Without the fetters of any relation to classical astronomy, theorists have had a field day trying to explain the observations. The requirements for a theory in these circumstances are that it account for the observations and be self-consistent. Plausibility is not necessarily a constraint because gamma-ray bursts represent a new and unprecedented phenomenon. At a meeting shortly after their discovery, Mal Ruderman who was giving the review talk on gamma-ray bursts announced that it was easier to give a list of the people who had not presented a theory of gamma-ray bursts than it was to give a list of those who had. He showed a slide consisting of one name, Jerry Ostriker who, for whatever reason, had not jumped on the gamma-ray burst bandwagon.

Theories have ranged from black hole collapse to "relativistic bb's." The latter were supposed to be little grains of dust accelerated to near the speed of light and then arriving at the Solar System to crash energetically into the solar wind. Remember all the billion pulsars that have died in the Galaxy? One of the first theories, and one that generated more than a few chuckles, postulated that gamma-ray bursts were generated by comets falling onto those neutron stars. One of the little-known but supportive ideas of this hypothesis is that clouds of comets may very well spread nearly from one star to another. Space may be filled with comets, and the chance that one of them would occasionally fall onto one of those billions of neutron stars is not so low.

The argument that swayed some people into taking this comet idea more seriously is the problem of generating gamma rays at all with a neutron star. The problem is related to the Eddington limit (Chapter 2). If energy is released on the surface of a neutron star, the material expands and cools in response to the radiation pressure. Under normal circumstances, such matter can get hot enough to emit X-rays, as we have seen in Chapter 8, but not hot enough to emit the more energetic gamma rays. The importance of the impact picture is that the material arrives in a lump and is compressed much more than would be either a dribble of gas or material just sitting on the surface. The effect might be enhanced if the infalling matter were a rock, so asteroids as well as comets have been considered. After a hiatus of a number of years, a similar idea was still around in 1998, although it has sunk under the weight of recent results.

There is a benefit to allowing the imagination of the theorists to run the bounds of the known data. What was really needed were more data so that theory and observation could march hand in hand in some fruitful direction.

2.2. The Revolution

All this changed with the launch of a Dutch-Italian X-ray satellite, *BeppoSAX* on April 30, 1996. This wonderful name derives from the nickname of a pioneering Italian physicist and X-ray astronomer, Giuseppe Occhialini, known as Beppo to friends and colleagues, with the appendage for X-ray satellite in Italian, Satellite per Astronomia a Raggi X. *BeppoSAX* was designed to look everywhere on the sky for the weaker X-ray signal that characterizes gamma-ray bursts and to give a first coarse location, more accurate than BATSE provided. The key innovation for *BeppoSAX* was a second instrument that can be brought to focus by quickly slewing the satellite in an attempt to rapidly find the X-ray flare from the gamma-ray burst and to provide a much more accurate location, with an uncertainty of a few minutes of arc. At that point, ground-based optical telescopes can be brought to bear to search the much smaller location to see if there is any optical component. All this was a bit of a gamble. If the whole gamma-ray burst phenomenon in lower-energy X-rays and in the optical faded in the tens of seconds that characterized the gamma-ray bursts themselves, then there would be no time to slew the satellite, a process that would take at least hours, never mind time to obtain optical images, a process that might take a day (or night) even in the best of circumstances.

Another chapter of this story is worth telling if only to recognize the great effort and ingenuity that goes into the scientific enterprise that sometimes fails to pay off. At a meeting on gamma-ray bursts in Santa Cruz in the 1970s, the attendees recognized that studies of gamma-ray bursts were stymied by the lack of observations at other wavelengths. A project was born to design a satellite that would contain a gamma-ray detector, but also ultraviolet and optical detectors to look in the same direction and hence to get simultaneous information on the burst at other wavelengths. The project was named *HETE* for *High-Energy Transient Explorer* and the arduous process of design began. It won NASA competitions to build and launch and suffered the inevitable delays. *HETE* was finally scheduled to launch on November 4, 1996, a date that would have put it in competition with *BeppoSAX*. The Pegasus rocket that carried *HETE* failed and the satellite was destroyed. That opened the way for *BeppoSAX*. To their credit, the *HETE* team regrouped, took the plans and spare parts, and built a new satellite. *HETE 2* is scheduled to launch in May 2000 and will be a valuable tool for the study of gamma-ray bursts.

BeppoSAX scored its coup on February 28, 1997, when it localized a burst sufficiently well that an optical follow-up was feasible. The result was the discovery of the first optical counterpart by a team led by Dutch astronomer Jan Van Paradijs. The fashion has been to label *BeppoSAX* gamma-ray bursts by the year and day that they were discovered. So far, no two *BeppoSAX* events have been discovered on the same day to mess up this scheme. The breakthrough gamma-ray burst was thus named GRB 970228. Two months later, in early May, *BeppoSAX* found another event, GRB 970508, enabling another optical identification. In this case, absorption lines of matter in front of this source prove that the source is at a cosmological distance, of order 1 billion light years or greater. In December of 1997, yet another optical counterpart was discovered associated with GRB 971214. After the gamma-ray burst faded, a faint galaxy was revealed. The red shift of this galaxy was immense, with the wavelength of the detected light shifted by more than a factor of 3 from its natural wavelength. This galaxy is estimated to be 12 billion light years away. If GRB 971214 is radiating equally into all directions and hence following the basic inverse-square law for apparent brightness (Section 1.2), then estimating the distance from the red shift (and adopting specific values of the cosmological parameters) implies that the energy of this source is fantastically large. More energy is required than the entire collapse and neutrino energy of a supernova and more than even most exotic theories of colliding neutron stars and black holes can support.

This story is still rapidly unfolding, and even GRB 971214 is not the record. That belongs to the first burst localized by *BeppoSAX* in 1999, GRB 990123. This burst brings in yet another interesting chapter in the saga. Many people realized that if an optical counterpart were ever to be seen, then an especially rapid response was needed. A special email notice system run by Scott Barthelmy and his colleagues at the NASA Goddard Space Flight Center in Maryland was set up. Even more extreme, some people began to wear beepers that were triggered electronically by a signal from a satellite, BATSE or *BeppoSAX*, so that they get buzzed the instant (allowing for the finite travel time of light and relay switches) a gamma-ray burst is detected. One of the things that this rapid response allowed was communication with automatically controlled robotic telescopes that would very quickly swivel to look for an optical counterpart, perhaps in the time frame of the original gamma-ray burst. This was the mission of *ROTSE*, the *Robotic Optical Transient Search Experiment*.

ROTSE is a small telescope situated at the Los Alamos National Laboratory. It was designed and operated by astronomers at the University of Michigan, Los Alamos, and the Lawrence Livermore National Laboratory. *ROTSE* was constructed to receive signals directly from the satellites that detect gamma-ray bursts and then to rapidly swivel and look at the location of a gamma-ray burst. *ROTSE* is not very sensitive as telescopes go because it has only four wide-angle camera lenses, but it can see a fairly large portion of the sky at one time to look for variable sources. The advantage is that it is quick! Quickness does not count if the weather does not cooperate or if the discovered gamma-ray burst is only visible from the southern hemisphere or if it is "up" in the north during daylight hours. This was the tale for the first number of *BeppoSAX* bursts. *ROTSE* did have a clean shot at some bursts, but it did not see anything.

Finally, on January 23, 1999, everything came together, and *ROTSE* scored its first detection of a gamma-ray burst. *ROTSE* detected the immediate optical counterpart of GRB 990123. The results were dramatic. *ROTSE* saw a burst rise in about 10 seconds to ninth magnitude and then fade over the next minute or so. This peak apparent brightness was only about a factor of 10 dimmer than can be seen with the naked eye! Associated work on this gamma-ray burst revealed it to be at yet another immense distance. This makes GRB 990123 the intrinsically brightest optical event ever recorded in scientific history. Ho hum, another record for gamma-ray bursts. Actually, there is nothing to be blasé about here. If radiated uniformly in all directions, the implied peak optical luminosity of GRB 990123 was equivalent to ten million supernovae or ten thousand very bright quasars. This optical burst did not last long, but its intensity was very impressive.

Most of the energy emitted by GRB 990123 was in the gamma-ray range. Here again, GRB 990123 set a record. The detected gamma-ray intensity was among the strongest ever seen at the Earth. At the distance observed, the total energy in gamma rays was ten times higher than the previous record-setters like GRB 971214. If this gamma-ray energy poured out equally in all directions, the energy involved was equivalent to the complete annihilation of 2 solar masses of matter! One runs out of exclamation points.

These optical counterparts of the cosmic gamma-ray bursts have thus revolutionized the field and proven the power of focusing optical astronomy on this decades-old problem. They have opened a new era in the study of gamma-ray bursts that promises not only rapid progress in understanding the bursts themselves, but also their use to explore the nature of the Universe at great distances.

The emission witnessed in the X-rays by *BeppoSAX*, in the optical by ground-based telescopes and in the radio by radio telescopes, was discovered to last much longer than the original gamma-ray burst. Rather than tens of seconds, the X-rays last for days, and the optical and radio can stay above limits of detectability for weeks or months. This delayed emission of energy has been termed the *afterglow* of the gamma-ray burst. The general interpretation is that the process that energizes the event, whatever that process is, sends a powerful explosion out into the interstellar gas surrounding the event. The explosion generates a strong shock wave that moves at very nearly the speed of light. The interaction of this shock wave with the interstellar gas can produce gamma rays, X-rays, optical emission, and radio emission in appropriate circumstances. The general process leading to this afterglow is called a *relativistic blast wave*. Models based on this process have been successful in accounting for many of the observations of the afterglow, including the spectrum of the radiation and the rate of decay that tends to drop off as one over the time since the original gamma-ray burst. If you wait twice as long, the glow is half as bright.

Many new pieces to the puzzle of gamma-ray bursts have been put on the table since the first *BeppoSAX* discovery. Many fundamental mysteries still remain. Despite the new information and the growing understanding of the afterglow phase, we still do not know the basic mechanism that releases the energy and converts so much of the energy into gamma rays. Some of the *BeppoSAX* events show optical afterglows, but others apparently do not. Most of the afterglows decay so that the power fades inversely with time, but some decay more rapidly. In a real sense, the field is just beginning and is likely to explode with activity.

2.3. The Shape of Things

One of the issues to be confronted in the study of gamma-ray bursts is the manner in which the energy is released into the surroundings. There are a number of tightly intertwined issues here. Theoretical models of relativistic blast waves and the afterglow demand that a shock wave moves out from the source at speeds very close to the speed of light. To do this, the flow of energy must carry along with it very few ordinary particles, protons, or more generally baryons. Too many of these particles of ordinary matter would slow the shock wave down so that it could not propagate with the deduced speeds. That is one thing that must distinguish an ordinary supernova and a gamma-ray burst. Both events have roughly the same amount of energy, but supernovae put their energy into moving a lot of ordinary matter at high, but not relativistic speeds. Gamma-ray bursts must put as much or more energy into a very small amount of mass.

Given the expansion at nearly the speed of light, a number of issues arise that come from Einstein's special theory of relativity. When motion with respect to an observer is high, lengths are foreshortened, and times are constricted. A gamma-ray burst that takes a minute as observed at the Earth may have spread over a region the size of the Solar System at its origin. An event that takes several months to play out in the host galaxy of the gamma-ray burst may take only hours or days as observed at Earth. In particular, it may take many months for the relativistic shock wave to expand out from the source of energy, pile up mass in the interstellar medium, and slow to ordinary speeds. An observer on Earth would see all this playing out in a day or so. Turned around, when we see a gamma-ray burst afterglow fading over a few days, it might have taken months in a far galaxy.

Another interesting effect is that, if a source of radiation moves toward an observer at a high speed, the radiation is thrown in the direction of the observer. This "beaming" can make the radiation seem brighter than it would otherwise be. In addition, if the source of energy is moving toward the observer, there is a very large blue shift, a "boost" of the energy of each photon that is detected. This can again make the source look brighter.

Such issues arise in trying to determine how bright a given gamma-ray burst really is and how much energy it emits. Even if the energy from a gamma-ray burst is emitted equally in all directions, it will be beamed and boosted and look brighter for a shorter time to an observer standing still on the Earth compared to an observer at the same distance who moved with the velocity of the shock. Trying to figure out how bright a given gamma-ray burst "really" is in its own rest frame is a rather tricky business that requires an understanding of just how the boosting and beaming is working.

One can get a measure of the total energy emitted in the radiation independent of the beaming and boosting if the energy is emitted equally in all directions. The procedure is to add up all the energy received at Earth over the course of the burst event. That energy might have been emitted over a different time span in the frame of the explosion, but all the energy is all the energy, and it must all go somewhere eventually. If one assumes it goes off equally in all directions and corrects for the fact that things look dimmer by the inverse square of the distance (plus perhaps some corrections for cosmological warping), then the total energy in radiation of the explosion can be determined. For the *BeppoSAX* events for which there is a measure of the red shift and hence the distance, the results are imposing, as mentioned earlier. For the event at the largest distance, 12 billion light years, the energy is comparable to the entire flow of neutrinos from a supernova, a huge amount of energy, and for GRB 990123 it is 10 times the neutrino energy of a supernova.

There is an important caveat to this method of measuring energy. If the flow of energy does not come out equally in all directions, if it is collimated in some way, if it flows out in a jet, then less total energy is required for a given observed burst, just in proportion to the amount of collimation, as shown in Figure 11.4. If the energy flows only into 10 percent of all available directions, then a given energy received on Earth requires only 10 percent as much total energy at the source. If the energy flows in a jet filling only 1 percent of the area around the source, then the energy at the source is only 1 percent of that deduced from the assumption that equal energy goes in all directions.

This collimation effect is not a fantasy. It is almost the rule rather than the exception. We see collimated flows from the Sun, protostars, planetary nebulae, binary black holes, and quasars. If the energy of a gamma-ray burst comes out in a collimated relativistic blast wave in only certain directions, then one must be careful in making estimates of luminosities and energies.

An example of this phenomenon is the "blazars." Blazars are a certain subclass of quasars that are especially bright and highly variable. The common interpretation is that in these objects we happen to be looking right down the nozzle of a jet of matter ejected at nearly the speed of light. By the accident of the Earth's position in the beam, we see an especially bright source of radiation because of the beaming and boosting associated with the rapid motion toward us. We also see especially rapid time variability of the radiation that is thought to be associated with the shrinkage of time due to the relativistic motion. No one suggests that this energy is flowing out equally in all directions, thus requiring unprecedented amounts of energy, even for quasars. Rather it is assumed that, if we happened to observe the same object from the side, it would resemble an "ordinary" quasar. Understanding whether, how, and how much gamma-ray bursts are collimated is one of the key tasks facing the field.

At this writing, *BeppoSAX* has discovered over two dozen gamma-ray bursts that have had observable X-ray afterglows. Some of these have produced evidence that the energy flow is, indeed, strongly collimated. The important evidence is that, even though some of the afterglows fade roughly inversely with time as expected for spherical relativistic blast waves, a few have been observed to decline more rapidly. One possible explanation is to invoke a jetlike, rather than spherical flow. A critical difference between a jet and a spherical blast wave is that, when it slows down, a jet can expand sideways. This sideways expansion can tap the energy of the jet and cause more rapid

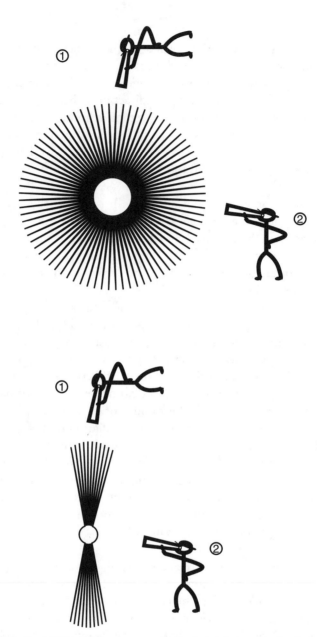

Figure 11.4. (top) If the energy in a gamma-ray burst flows out equally in all directions, then it does not make any difference where the observer is. All observers at the same distance will see the same brightness and deduce the same energy. (bottom) If the energy is collimated into a jet, however, the observer (#1) who looks down the jet will see a much higher luminosity than the observer (#2) looking from the side. If observer #1 assumes that the energy is emitted equally in all directions, he will deduce too large a total energy for the event.

cooling and deceleration and hence a more rapid rate of decline of radiation output. Several of the brightest gamma-ray bursts have the most clear evidence for this behavior, and quantitative analysis is consistent with their being collimated to only 1 percent of the sky, or even less. This means that the energy is reduced by a factor of 100 or more, and that they must be 100 times more common. This strong collimation has been specifically attributed to GRB 990123, bringing its energy down from a mind-boggling level equivalent to the expansion energy of three thousand supernovae to about 10 percent of the total collapse energy of a neutron star, only ten times the expansion energy of a normal supernova.

Although there is a growing belief that many if not most gamma-rays bursts and their afterglows are jetlike, there are other explanations for the rapid decline. If gamma-ray bursts arise somehow in massive stars, then they should be surrounded by the matter blown off in a stellar wind. A spherical blast wave will collide with this wind and slow much faster than if it only interacted with the dilute matter of the interstellar medium. This interaction can also account for the rapid declines seen in some afterglows. There is, of course, nothing to prevent a jet from colliding with a wind, and if the source of gamma rays pumps out energy for a prolonged time, the tendency for the power to decline can be overcome, so there are lots of complications to be pursued and understood.

2.4. Supernovae and Gamma-Ray Bursts

Despite the lack of understanding of the source of energy of gamma-ray bursts, they are widely thought to be at least vaguely related to supernovae. Many models require objects that are produced in gravitational collapse – neutron stars or black holes – to collide or merge or spin or swallow matter in some dramatic way. In the onrush of events that followed from the *BeppoSAX* discoveries, another surprise made the relation of gamma-ray bursts and supernovae explicit.

On April 25, 1998, *BeppoSAX* discovered a gamma-ray burst, GRB 980425, of otherwise ordinary properties in terms of its apparent brightness, energy, and time scale. *BeppoSAX* then swung to bring its fine position sensor X-ray detector into position and detected a couple of X-ray sources, one of which diminished in time and one of which seemed to be constant. A day later, optical astronomers caught up and found a strongly variable object. This object was not, however, the afterglow that one had quickly learned to expect. It was, rather, a supernova, one of rather strange properties. The supernova, SN 1998bw, was not exactly at the position of either of the two X-ray sources first reported by *BeppoSAX*. This raised some question about the association of SN 1998bw with GRB 980425. In the next few months, the *BeppoSAX* team recalibrated the positions of the X-ray sources they detected. The source that was at first observed to vary was determined to be much too far from SN 1998bw to be associated. The other source, at first thought to be constant, was shifted so that an association with SN 1998bw could not be ruled out. Then this source was discovered to be variable, if only slightly. This has left the issue of the association of SN 1998bw with the *BeppoSAX* X-ray sources somewhat befuddled. One must be wary of other sources of variable X-ray emission, such as active galactic nuclei that could accidentally fall near the supernova, but an association of one of the X-ray sources with SN 1998bw cannot be ruled out.

A few days after the detection of the SN 1998bw, radio astronomers found a very bright radio source. This radio source is precisely at the position of SN 1998bw, so there is no question of their association. Analysis of the radio data showed that the radio source is brighter than can be easily explained without expansion of a shock wave at nearly the speed of light. Independent of the gamma-ray burst, SN 1998bw clearly produced a relativistic blast wave. All this evidence taken together suggests that SN 1998bw and the gamma-ray burst GRB 980425 are one and the same thing. The likelihood of finding both GRB 980425 and SN 1998bw in the same part of the sky in the brief interval of time when they erupted is very low, so many astronomers think the connection must be real. In particular, even though gamma-ray astronomers tend to be leery of the association, supernova mavens have embraced it with full passion.

Observations of SN 1998bw and its host galaxy showed that it was at a distance of about 40 million parsecs or about 120 million light years. That is a great distance, but far less than the record-setting 12 billion light years of GRB 971214. At 40 million parsecs, the total energy in the gamma-ray burst is deduced to be much less than that of the most powerful gamma-ray bursts, by a factor of about 1 million. On the other hand, at the same distance, SN 1998bw is exceptionally bright for a supernova. Both of these results are puzzles that must be assimilated in the ongoing attempt to understand gamma-ray bursts.

Although it is a step along an esthetically ugly path, one idea that emerges from this new event is that there are at least two kinds of gamma-ray bursts, one of very high energy seen at cosmological distances and one of lower energy seen relatively nearby. This is an uncomfortable hypothesis given that the gamma-ray properties of GRB 980425 were seemingly unexceptional. The similar nature of far-away energetic and nearby lower-energy gamma-ray bursts may arise because any physical events that can emit gamma-rays will have certain properties in common whether the total energy involved is high or low, but this remains to be shown.

SN 1998bw brings its own set of questions. The early spectra were unlike any other supernovae we have discussed, Type Ia, Ib, Ic, or II. With hindsight, there were a few other supernovae, SN 1997ef is a conspicuous example, that did bear some resemblance to SN 1998bw, so there may be some precedent. As it evolved, SN 1998bw looked more and more like a Type Ic with no evidence for hydrogen or helium. It certainly did not look like either a Type II or a Type Ia.

The first models of the light curve and spectra have assumed that SN 1998bw resulted from core collapse, and that enough radioactive nickel was produced to power the peak of the light curve. Because SN 1998bw was about as bright as a Type Ia (even though the spectrum is completely different), a comparable amount of nickel is required, about 0.7 solar mass. Basic spherically symmetric models can produce this amount of nickel in a core-collapse explosion by shocking silicon layers, but they are extreme. Models that make this much nickel and that produce the observed light curve and spectra at some level of agreement (not perfect in the first models) require an exploding carbon/oxygen core of about 10 solar masses and an energy of expansion of the matter of more than ten times that normally associated with supernovae. These models suggest that SN 1998bw was a "super" Type Ic, and the term "hypernova" has been adopted in some circles.

If this interpretation of the observations is right, then SN 1998bw is an exceptional event. It must involve processes unlike all the other supernovae discussed in this book. Speculations run toward the creation of a black hole in the explosion and subsequent accretion of matter to fuel the large energy release. Another possibility is that the explosion is asymmetric so that the explosion looks brighter at some viewing angles than others. All these possibilities need to be explored.

My colleague Lifan Wang and I have proposed a picture based on the nature of GRB 980425 and SN 1998bw that might unite supernovae and gamma-ray bursts. The idea is to see how far one can go with using only relatively ordinary supernovae to produce gamma-ray bursts. The argument is that all gravitational collapse events produce strong magnetic jets that punch out through the axes of the surrounding carbon/oxygen core. In ordinary Type II supernovae, the outer hydrogen layers would stop these jets. In Type Ic or Type Ib, the jet could escape into interstellar space making the gamma-ray burst. This is shown schematically in Figure 11.5.

In this picture, there are two components to the gamma-ray emission, one that radiates more or less equally in all directions with the energy seen in GRB 980425/SN 1998bw, about one thousand times less than a standard supernova expansion energy, and one component that is highly collimated in a relativistic jet containing perhaps 10 percent of the total supernova energy. The lower energy component could be seen if the explosion occurred relatively nearby, at 100 million light years, but would not be detectable with current instruments if the same event were at truly cosmological distances. The other gamma-ray component emerges in the jet so that all the gamma-ray energy contained in it is collimated to flow in a narrow angle. In this way, only some fraction of the supernova energy is required to be channeled into gamma rays. By this argument, the huge energies deduced for the very distant gamma-ray bursts is an artifact of assuming that equal energy is emitted in all directions, rather than being confined to the direction of the jet, as in the blazar picture described earlier. To reduce the required energy from the amount deduced in an "all directions" picture to some fraction of a supernova energy, the jet must be tightly collimated. The area of its cross section must be only one part in a thousand of the area surrounding the burst source. This is about the amount of collimation seen in typical jets from active galaxies and comparable to the collimation deduced from some of the gamma-ray burst afterglows by their rapid decline, so it is not beyond the bounds of credibility. Whether it is produced in a real supernova is another story that will require intensive investigation.

If the jet moves at nearly the speed of light, the gamma rays will be blue-shifted and beamed strongly in one direction. This component could, in principle, be seen at cosmological distances if the jet happens to be pointed right at the Earth. Most of the jets will not be pointed at the Earth, so this picture requires many more gamma-ray burst events that are not pointed at the Earth to account for the few that are. If the collimation is to one part in a thousand, then there must be one thousand jets not pointed at the Earth for every one that is.

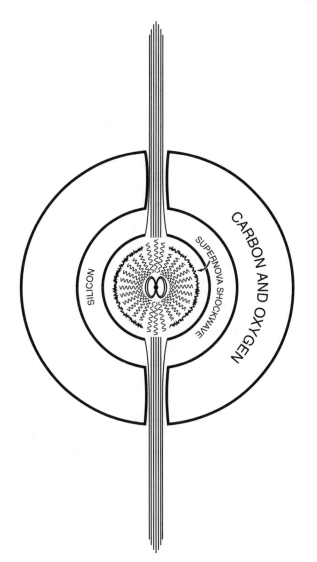

Figure 11.5. A possible schematic model for a gamma-ray burst in a Type Ic supernovae. The evolution of a massive star that has lost its envelope proceeds to the formation of an iron core that in turn collapses to form a neutron star. The formation of the rotating, magnetized neutron star, perhaps a highly magnetized magnetar (see Chapter 8) sends out jets of material that punch holes in the core as the outer layers of matter hover, waiting to collapse or explode. The newly formed pulsar then fills the inner cavity where the iron core had been with intense electromagnetic radiation. This radiation floods out the wounds in the core punched by the earlier jets and forms the gamma-ray burst. Such a highly collimated gamma-ray burst could only be seen at special angles looking right down the beam. This configuration could, in principle, yield a gamma-ray burst that could be seen across the Universe but requires only a portion of the supernova energy. A small fraction of the gamma rays could be radiated sideways, making the weaker gamma-ray burst associated with SN 1998bw. The rotational energy of the pulsar might enhance the explosion of the supernova itself.

The required rate is roughly that for normal supernovae, approximately one per few hundred years per bright galaxy, giving a crude concordance to the argument.

There is some weak statistical evidence that some Type Ic supernovae and related events are associated with gamma-ray bursts, but there are many loose ends to this picture. How would the jet form and would it emit gamma rays and the other products of the afterglow in the right way? What is the mechanism of the lower-energy gamma-ray component? Why should the weaker gamma-ray components that emerge in all directions have the same observed average properties as the emission from a beamed relativistic jet? The lower-energy component should play out in ordinary time, but the jet component should suffer a strong Einsteinian time compression. Why do they both last about a minute? Should every Type Ic show evidence for the jet that could long outlive the supernova? Could SN 1998bw possibly be related to other more normal Type Ic despite the evidence for very high ejecta masses and high explosion energies? There is evidence that "ordinary" Type Ic are not spherically symmetric but that they produce explosions that are distorted in a significant way. This already suggests that the core collapse that generates Type Ic, and presumably all other collapse-driven supernovae, is strongly asymmetric, perhaps involving jets shooting out the rotation or magnetic axes. Could such events be brighter in one direction than another, accounting for the apparent excess luminosity of SN 1998bw? Is the evidence for non-spherical ejection in Type Ic related to the jets necessary to make gamma-ray bursts that can be seen across the Universe? If soft gamma-ray repeaters require magnetars (Chapter 8), neutron stars with superstong magnetic fields, what happens at their birth? Are magnetars born in binary star systems that spin up the star? Do they generate a jet when they form? Do they resemble Type Ic supernovae or SN 1998bw? These are all questions that need to be addressed.

Like Type Ic, SN 1998bw showed signs of asymmetry, evidence that the flow of ejected matter departs rather strongly from spherical symmetry. This evidence was ignored by the first spherically symmetric hypernova models that require unprecedented amounts of energy to provide the supernova luminosity. Peter Höflich, Lifan Wang, and I have considered models that are distorted by a sufficient amount to account for the asymmetries in Type Ic supernovae and in SN 1998bw itself. Preliminary models show that, if the ejecta are in the shape of a fat pancake, they will be considerably brighter if viewed from the top of the pancake compared to the edge, by about a factor of 2. These models have the potential, at least, of accounting for the observed optical properties of SN 1998bw with "normal" amounts of energy and ejected nickel mass. Whether such models, or the hypernova models for that matter, can account for the gamma-ray properties remains to be seen.

An important step in addressing these issues is a calculation done by Alexei Khokhlov at the Naval Research Laboratory, me, and our colleagues. This calculation glossed over a number of complications that need to be addressed more deeply but assumed that a newly formed neutron star could launch jets along the rotation axes in about a second while the outer parts of the star hovered, waiting to be blasted into space or to collapse into a black hole depending on the outcome of the collapse. To correspond to a Type Ib or Ic supernova or to SN 1998bw, the hydrogen envelope of a massive star model was omitted, and only the core of helium and heavier elements was retained. The jets penetrate to the surface of the helium core in about 6 seconds. As they propagate, the jets drive bow shocks that blow sideways as well as forward. The result is that shortly after the jets penetrate the surface, the sideways shocks converge and eject the matter at the equator and all directions in between. If the star has no hydrogen envelope, as assumed by Khokhlov, then the final result is two jets of matter along the axes and a strongly asymmetric, pancake-like explosion otherwise, just as illustrated schematically in Figure 11.5. This explosion is driven entirely by the jets. The stalled shock and the neutrinos described in Chapter 6 play no role. Further study may show that both jets and neutrinos are necessary in varying degrees, and this calculation does not directly address the production of gamma-ray bursts, but it is a sterling example of how new insights from the study of gamma-ray bursts are driving new ideas in supernova research and vice versa. If jets are a critical part of the explosion in many if not all core collapse events, then many issues such as nucleosynthesis and the production of black holes must be reconsidered.

The question of the connection of supernovae and gamma-ray bursts was fueled by developments in the spring and summer of 1999. One gamma-ray burst from 1998 was later found by Shri Kulkarni, Josh Bloom, and their colleagues at Caltech to show evidence for a brightening about 3 weeks after the gamma-ray burst that interrupted the otherwise rather rapid (and hence from a jet?) decline of the afterglow. This apparent new source of light was roughly consistent with the addition of the light from a "SN 1998bw-like" event that reached peak about 3 weeks after the gamma-ray burst, a reasonable time for a supernova to have attained maximum light output after its initiation. After this discovery, the original afterglow event, GRB 970228, was also reanalyzed; evidence for a "SN 1998bw-like" brightening was deduced, and similar arguments were advanced for one or two more events. The statement that the excess light in addition to the afterglow might be similar to SN 1998bw is carefully couched to avoid any specific commitment to whether SN 1998bw was itself a more or less "normal" supernova, or the first example of a new class of hypernova events that are solely related to gamma-ray bursts. Time will tell.

2.5. The Possibilities

These years of living dangerously after the first *BeppoSAX* discovery have left a large range of possibilities for the origin of gamma-ray bursts. They cannot all be true, but it will take a lot of work and ingenuity to show which ideas will survive.

The idea that gamma-ray bursts are not at cosmological distances, but relatively nearby, has essentially died, but it lingers in the back of some people's minds. People of this bent wonder whether the apparent association of gamma-ray bursts such as GRB 971214 with a very distant galaxy is an accidental alignment, the eternal bugaboo of astronomers who peer out into the huge volume of space. If you look far enough, so this argument goes, you are bound to see a galaxy, so what? Estimates of the probability are about one in a thousand, so such an unhappy accident is unlikely, but not impossible.

The latest wrinkle in this area comes from an exciting new development in the study of soft gamma-ray repeaters (Chapter 8, Section 10). Gamma-ray bursts divide into two distinct categories, most last about 30 seconds, as described earlier. A significant minority, however, last less than a second; a few tenths of a second is typical. Another puzzle of gamma-ray burst research has been to understand this dichotomy in temporal behavior. Does this represent variations on a theme, or two distinct physical processes? A recent discovery, announced in September 1999, is that one of the soft gamma-ray repeaters produced a gamma-ray burst that is indistinguishable from the "short" gamma-ray bursts. Because the soft gamma-ray repeaters are magnetars in our Galaxy, this raises the question of whether or not all the short gamma-ray bursts are from neutron stars in our Galaxy. If this is so, their distribution should not be uniform on the sky because of the Sun's offset position from the center of the Galaxy. For technical reasons, *BeppoSAX* cannot respond to these short bursts, so everything that has been learned about gamma-ray bursts and their afterglows in the *BeppoSAX* era pertains only to the "long" gamma-ray bursts. HETE 2 will be able to detect and study the short bursts, now, more than ever, an important task.

Then there is still a plethora of models that address the gamma-ray burst energy issue head on. Some of these involve colliding neutron stars. That process has plenty of energy, maybe enough for the most extreme events if the energy emerges in a jet. Another principal issue is turning the energy into gamma rays and a relativistic blast wave that is not so overloaded with protons that it cannot move rapidly enough to make the burst or the afterglow. One possibility is that the neutron stars do not collide directly but interact through their strong magnetic fields. That way one can think about turning the pure magnetic energy into pure gamma-ray energy without getting the stuff of the neutron stars, those troublesome, slowing baryons, directly involved. Other models invoking neutron stars suggest that the powerful radiation from a newly born pulsar could result in a gamma-ray burst. Yet other models envision a disk of matter carrying a large magnetic field that accretes into a newly born, rapidly rotating black hole. If the accretion is fast enough and the magnetic field strong enough, large energies can be created. Stan Woosely of the University of California at Santa Cruz and his colleagues have argued that such energy can emerge in a jet that could also blow up the surrounding star as it moved out from the black hole. This mechanism might make a supernova-like explosion from a black hole as well as a gamma-ray burst. The energy can be scaled up with the mass of the black hole, but these models collimate the energy in a jet, so the total energy requirements would be correspondingly muted. The open issues are whether the right kinds of black holes form with the right accretion and the origin of the needed magnetic fields. Yet another version of this picture invokes the collapse of supermassive stars of perhaps more than ten thousand times the mass of the Sun. This process is speculated to occur in any case to form the giant black holes we believe reside in most large galaxies, if not all galaxies, and that give rise to quasars. Because the mass involved in the collapse can be very large, the energy released can also be very large, so that only a small fraction of the total needs to be converted to gamma rays.

All these pictures have a certain basic plausibility about them, given that we think our Universe is full of magnetic neutron stars and black holes of a range in mass from those of stars to those of galaxies. The devil is in the details. Having accounted for the energy, the first major requirement, can any of these models really account for gamma-ray bursts with the observed properties? All these models designed to give very high energy gamma-ray bursts at cosmological distances must also now confront GRB 980425 and SN 1998bw. How is it that a newly formed accreting black hole in the young Universe produces a gamma-ray burst with the same average observed properties as a relatively nearby, much less energetic, odd supernova?

These are the conundra that make astrophysics so exciting. Gamma-ray bursts will continue to provide all the stimulation an astrophysicist could want for some time to come. As better understanding of the gamma-rays bursts develops, so will a better understanding of the Universe on both stellar and cosmological scales. The gamma-ray bursts give us yet another means to look throughout the space and time of our visible Universe.

BATSE detects about one gamma-ray burst per day. If every one were like GRB 990123, there should be a bright optical flash for a minute or so once a day somewhere on the sky that would be easily visible with a decent pair of binoculars. A pair of binoculars allows you to see about one part in a thousand of the total sky. If you looked every night for 3 years running, you just might get lucky. Here comes backyard cosmology.

12

Black Holes, Worm Holes, and Beyond

The Frontiers

Black holes may form from stars, but they are vastly different from stars. One way to see this is to examine the intellectual frontiers to which research on black holes has led. There one finds mind-bending concepts of worm holes, time machines, multidimensional space, self-reproducing universes, and radical new notions of how to think of time and space under conditions where neither can exist.

TRISPATIOCENTRISM

"Egocentric." "Enthnocentric." A variety of words in the English language describe the tendency of people to get locked into a limited perspective. "Anthropocentric" is a favorite word in some circles of astronomy. It describes the tendency of scientists as well as *Star Trek* writers to conjure up alien life forms that are fundamentally similar to us, not just physically, but emotionally and socially, with our motivations, drives, and dreams. The *anthropic principle* – that the Universe is as it is because we exist – is a related idea. In the never-ending battle to expand our perspectives, I write this to call attention to the existence of another limited, rarely questioned, viewpoint that affects us all: *trispatiocentrism*. Trispatiocentrism is the attitude that the "normal" three-dimensional space of our direct perceptions is all there is and all that matters.

This word arose in my substantial writing component course at the University of Texas in Austin. We were exploring the nature of space and time with a particular emphasis on spaces of various dimensions. I wanted a word to connote the notion that our three-dimensional world view carries with it unrecognized restrictions. I came up with "trispatiocentric" and its obvious variations.

There is a serious scientific side to this. Some understanding of curved space is needed to picture how Einstein's theory of gravity works. To illustrate the basic ideas, gravitational physicists often have recourse to examples of curved two-dimensional spaces, the surfaces of spheres or of saddles or of donuts. In these examples, our familiar three-dimensional space surrounds the surface so that we can easily envisage the curvature. The trick is to try to perceive what the corresponding curvature of our own three-dimensional space is like. The goal is to understand the arcanae of Einstein's theory: black holes, worm holes, and, more recently, time machines. In this context, it is quite natural for a logical, if naive, mind to ask: if the surface of a sphere curves in a three-dimensional space, then must our three-dimensional space curve in some four-dimensional space?

For the nonnaive, these issues arise at the forefront of modern physics, the attempt to construct a "theory of everything." This theory will allow us to understand the raging singularities predicted to be at the centers of black holes and from which the Universe was born. Singularities represent the place where our current concepts of space and time, indeed all of physics, break down. The most successful current attempts to develop a new understanding of space and time are based on "string theory," where, to be self-consistent, the "strings" that constitute the fundamental elements of nature exist in a space of ten or more dimensions. Thus these developments have led physicists to ponder higher dimensions, albeit ones so tightly packed we cannot perceive them directly. They speak in terms of surfaces or membranes in a space of p-dimensions and call them "p-branes." Alas, I cannot resist pointing out that all this is not for pea brains like me. It is, however, the stuff that will push back the frontiers of knowledge and along the way help to resolve famous wagers made by Stephen Hawking concerning the nature of space and time.

In our course, we read the classic old tale *Flatland* by Edwin Abbott. Here we meet the Monarch of Line Land who, in blissful ignorance, suffers his monospatiocentrism. The hero of *Flatland* is a simple square who is ripped, to his ultimate chagrin, from his bispatiocentric world view by a visitor from a three-dimensional Universe we would recognize.

Abbott, Einstein, and the work of string theorists would have us ponder a fundamental verity. We are gripped in a trispatiocentrism we rarely stop to recognize and even more rarely take the time to ponder. Why does our familiar space have three dimensions, no more, no less? Is the notion that this space is natural or even unique as archaic and limited as the notions that the Sun goes around the Earth or that the Solar System is in the center of the Universe? Is Heaven not "up" in a literal sense but in a higher dimension we cannot perceive? If so, what of Hell? When Captains Kirk, Picard, or Janeway are transported to a different dimension, why is it always so boringly and trispatiocentrically of a familiar number of dimensions? We are trapped in this three-dimensional world of our direct perceptions and scarcely know it.

Is it possible that space can be prized open with "exotic matter" leading to worm holes that reconnect time and space? Are the ten or eleven dimensional spaces of string theory the first hint of the "subspace" of *Star Trek*? The work of physicists on the vanguard of knowledge provides the first glimpses of what may exist beyond or without.

The hero of *Flatland* was imprisoned for attempting to challenge the bispatiocentrism of his peers. My students seem to have the same dismal expectations for any departures from societal norms. The stories they wrote for class of other-dimensional worlds suggested that society is likely to find unwelcome any assault on cherished "centrisms." With their stories as a guide, I should expect with this contribution to be summarily institutionalized, incarcerated, or executed. Nevertheless. the truth must be exposed.

Citizens of this three-dimensional Universe unite! You have nothing to lose but your branes!

1. Worm Holes

"Time is the fire in which we all burn," says a character in a *Star Trek* movie. This quote captures the hold that time has on our imaginations. Time, especially the fascinating and philosophically thorny issue of time travel, has been a common topic of science fiction since the classic story of H. G. Wells. The ability to manipulate time remains beyond our grasp, but physicists have conducted a remarkable exploration of time in the last decade that once again brings us to the frontiers of physics.

Separation of time from space has been a part of physical thinking since at least the era of Galileo. The equations physicists use to describe nature are symmetric in time. They do not differentiate time running forward from time running backward. A movie of dust particles floating in a sunbeam would look essentially the same run forward or backward. If the projectionist ran a regular film backward, you would notice immediately. Where does the difference, the "arrow of time," arise? Why is it that we age from teen age to middle age, but not the other way around? Is that progression immutable?

This particular attack on time travel arose from a work of science fiction. In the original version of his novel, *Contact*, Carl Sagan wrote of a mode of interstellar travel created by an ancient extraterrestrial civilization. He had in mind that his passageway was a black hole where you could fly into the event horizon and emerge – elsewhere. Sagan sent the draft of the book to Kip Thorne, a physicist at Caltech, and one of the world's experts on black holes. Thorne has written his personal version of this story in the book *Black Holes and Time Warps: Einstein's Outrageous Legacy*. Thorne realized that what Sagan proposed would not work. Thorne proposed a solution with both different physics and more imagination!

Einstein's equations for a black hole do describe a passage between two universes or between two parts of the same universe: a structure called an *Einstein-Rosen bridge*, or in more casual language, a *worm hole*. This is yet another phrase invented by the word-master physicist, John A. Wheeler. Black hole experts have known for decades that the apparent worm hole in the standard black-hole solution represents only a single moment in time. Just before or just after that instant, there is no passage, only the terrible maw of the singularity, waiting to destroy anything that passed into the event horizon. For an intrepid explorer who tried to race through the worm hole in the instant it opened at anything less than the speed of light, the worm hole would snap shut. The explorer would be trapped and pulled into the singularity.

Thorne realized with further reflection that there might be another approach. Suppose, he reasoned, you were dealing with a very advanced civilization that could engineer anything that was not absolutely forbidden by the laws of physics. Thorne devised a solution that was bizarre and unlikely, but could not be ruled out by the currently known laws of physics. His solution involved what he came to call *exotic matter*.

Ordinary matter has a finite energy and exerts a finite pressure and creates a normal, pulling, gravitational field. One can envisage mathematically, however, matter that has a negative energy, that exerts a negative pressure, like the tension in a rubber band. For exotic matter, this tension is at such an extreme level that the tension energy is greater than the rest mass energy, $E = Mc^2$, of the rubber band. Such material has the property one would label "antigravity." Whereas ordinary matter pushes outward with pressure and pulls inward with gravity, exotic matter pulls inward with its tension and pushes outward with its gravity.

Remarkably, related stuff has become a prominent topic in cosmology, as mentioned in Chapter 11. Cosmologists describe an inflationary stage occurring in the split seconds after the big bang, in which the Universe underwent a rapid expansion that led to its current size and smoothness. The condition that is hypothesized to cause inflation is some form of negative energy field that would have a negative pressure that pushed against normal gravity resulting in rapid expansion. After a brief interval of hyperexpansion, this field is presumed to decay away leaving what we regard today as the normal vacuum with its small but nonzero quantum vacuum energy. Another version of these ideas arises in the context of the cosmological constant discussed in Chapter 11. If the Universe is accelerating in its expansion, there must be something involved other than the matter in it and the irreducible quantum energy of the vacuum. One suggestion is yet another negative energy field, one invented to account for the acceleration of the Universe like the cosmological constant, but one that varies in time. This field has been termed *quintessence*. Thorne did not attempt to make the nature of exotic matter explicit. In the most general sense, however, the exotic matter needed to create worm holes would share some of the repulsive properties of the inflationary energy and quintessence.

Because it was not forbidden by physics, Thorne speculated that an advanced civilization could slather some of this exotic matter on a mortar board, pick up a trowel, and do something with it. Cleverly applied, the repulsive nature of the gravity of the exotic cement could hold open an Einstein-Rosen bridge indefinitely! Thorne had discovered, conceptually at least, a way to traverse through hyperspace from one place in the Galaxy to a very distant one in a short time. The result would effectively be faster-than-light travel through a worm hole, just the mechanism that Sagan wanted to further his plot. Sagan adopted Thorne's basic idea and described such a worm hole in the book that went to press. The movie was finally released in the summer of 1997, but, sadly, Carl Sagan succumbed to a rare disease and was not there to see how his vision of a worm hole translated to the screen.

Having passed the basic idea on to Sagan, Thorne remained deeply intrigued. He continued to work on the idea with students and together they published a number of papers showing that a proper arrangement of exotic matter could lead to a stable, permanent worm hole.

It is tempting to ask what a worm hole would look like. A worm hole would not necessarily look black, like a black hole, even though the outer structure of their space-time geometries were similar. A black hole has an event horizon from within which nothing can escape. By design, however, you can both see and travel through a worm hole. In its simplest form, a worm hole might appear spherical from the outside, that is, all approaches from all directions would look the same. If you travel through one, you would head straight toward the center of the spherical space. Without changing the direction of your propagation, you would eventually find yourself traveling away from the center, to emerge in another place.

A worm hole is not literally a tunnel in the normal sense with walls you could touch, but from inside a spherical worm hole, the perspective would be tunnel-like. You would be able to see light coming in from the normal space at either end of the worm hole. The view sideways, however, would seem oddly constricted. The space-time of the interior of a worm hole is highly curved. Light heading off in any direction "perpendicular" to the radius through the center of the worm hole would travel straight in the local space but end up back where it started, like a line drawn around the surface of a sphere, only in three-dimensional space. If you faced sideways in a worm hole, you could, in principle, see the back of your head. In practice, the light might be distorted and your view very fuzzy. The effect might look like a halo of light around you that differentiated the "sideways" direction from that straight through the center of the worm hole. Figure 12.1 shows how it might look to you as you shined a flashlight on the interior of the worm hole.

A common misconception is to confuse the tunnel-like aspects of a worm hole with the funnel-like diagram that physicists use to make a two-dimensional representation of the real three-dimensional space around a black hole or worm hole. In a two-dimensional embedding diagram, a circle in two-dimensional space is the analog of a sphere in three-dimensional space. The real curved space around a three-dimensional worm hole is represented in two dimensions by a stretched two-dimensional space that resembles a funnel, just as it was for a black hole, as discussed in Chapter 9. In this two-dimensional analog, you cannot travel through what we perceive to be the mouth

Figure 12.1. A flashlight beamed into a worm hole would shine out the other end, but one aimed sideways would illuminate the back of your head.

of the funnel. That is a third-dimensional hyperspace in the two-dimensional analog. You have to imagine crawling, spider-like, along the surface of the two-dimensional space to get the true meaning of the nature of that space and some feeling for the three-dimensional reality. A version of this two-dimensional analog of a worm hole is shown in Figure 12.2. The worm hole in Figure 12.2 connects two different parts of an open, saddle-shaped universe. One can also picture a worm hole cutting through a sphere in the two-dimensional analogy of a closed universe. It is more difficult to portray in an illustration, but worm holes can also provide such shortcuts in flat space. If they are properly designed, worm holes can, in principle, yield an arbitrarily short path between arbitrarily distant reaches of normal space in any sort of universe.

Recent movies and TV programs have been based on some of these modern notions of worm holes, but there is still a tendency to confuse the actual tunnel-like nature with the two-dimensional funnel-like

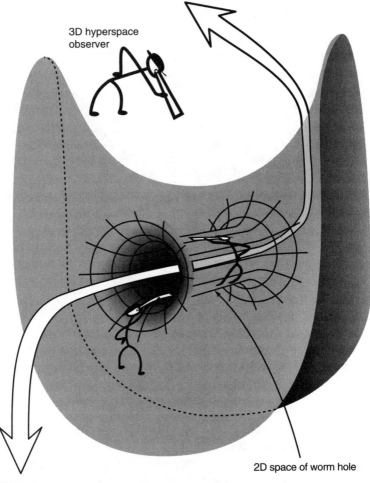

3D hyperspace
observer

2D space of worm hole

3D hyperspace
through hole

Figure 12.2. A two-dimensional worm hole giving a shortcut through an open saddle-shaped universe. In this representation, the three-dimensional space surrounding the Universe and threading the worm hole is a hyperspace that two-dimensional residents of the universe could not perceive. A two-dimensional denizen of the two-dimensional universe could approach this worm hole from any direction in 360 degrees and pass through the worm hole along the two-dimensional surface to emerge on the other side of the universe. An astronomer near the ''mouth'' of the worm hole could see a colleague within the worm hole, and vice versa. The astronomer within the worm hole could travel ''straight'' on a path at right angles to the way in or out and end up back where he started.

analog. In the first *Star Trek* movie, the *Enterprise* is captured in a worm hole when it jumps into warp drive too soon after leaving Earth. You can see stars through the sides of the worm hole. That is definitely wrong. Light from stars could come in the end of the worm hole the *Enterprise* entered, or it could come in through the opposite end toward which the ship is headed. Inside the worm hole, however, light is trapped by the severe curvature of the space. There is no literal tunnel wall; hence, Kirk and his crew cannot look out "sideways" through it.

The TV series *Babylon 5*, features a "constructed" worm hole, but its whirlpool-like nature is more reminiscent of the two-dimensional analogy than the proper manifestation in real space. In *Deep Space 9*, the worm hole can be approached from any direction and the tunnel-like interior is as close to "reality" as one can expect from graphic designers appealing to a TV audience. *Sliders* also does a pretty good job of capturing the spirit that the worm hole is basically spherical so the characters can enter and exit anywhere in three dimensions. The film *Stargate* and the TV program based on it show the worm hole portal to be a single flat, circular sheet. The characters enter and exit from only one side. That is Alice's looking glass, perhaps, but not well-rooted in this particular bit of science.

The classic worm hole is that in the movie *2001: A Space Odyssey*. The fact that the monolith orbiting Jupiter is a worm hole is a bit obscure, but that is what it is. In that film, the exterior of the worm hole is three-dimensional, but it is a flattened rectangle. Matt Visser of Washington University of St. Louis has designed a worm hole that looks much like that, with the exotic matter confined to struts along the boundaries of the rectangular body. In the movie version of *Contact*, the heroine is thrust into a worm hole by an alien-designed machine that opens the portal to the worm hole. The tunnel-like aspects are portrayed reasonably realistically, and there is an attempt to invoke the other amazing property of worm holes, the distortion of time.

2. Time Machines

If exotic matter, antigravity, and superluminal travel were not enough, there is even more to the worm hole story, and time is its essence. As they worked on the nature of worm holes, Thorne and his co-workers realized to their amazement that worm holes must also function as time machines. In this phase, Thorne was joined by Igor Novikov, then of Moscow, now at the University of Copenhagen, and his colleagues. A key aspect of the next stage of their thinking is what has been called the "twin paradox."

This conundrum arises already in the context of Einstein's special relativity. Einstein's theory shows unequivocally that a pair of twins moving at some velocity with respect to one another will each measure the other to be aging more slowly. The twin paradox apparently arises when one of the twins rockets out into space and then returns while the other remains at home. The motion is relative, but the twins cannot each be younger than the other. Is one twin younger, and, if so, which one? The resolution to the paradox is that the one that traveled will be younger. That traveler must have experienced a force, an acceleration upon turning around, and that makes all the difference. That is the answer when carefully analyzed with special relativity accounting for the acceleration that the traveling twin felt and the stay-at-home did not.

Thorne realized that you could do this experiment, again conceptually at least, with the two ends of a worm hole. Grab one end (gravitationally), and rocket it out and back. It will be absolutely younger than the end that was not accelerated. Novikov realized that the same result will arise by putting one end of a worm hole in empty space and the other near a gravitating body. General relativity says that time will flow more slowly in the gravity well. The end of the worm hole deep in the gravity would be younger than the end in deep space.

In either of these arrangements, you have a time machine! You can walk into one end of the worm hole and emerge in an earlier era. If you walk to the first end of the worm hole though the exterior space, time passes, and you age normally. You could meet your younger self before you entered the hole! Because this is science, not fiction, there are limits. You cannot exit before the worm hole time machine was created, so you cannot travel arbitrarily far back in time.

Time travel, including that invited by worm-hole time machines, leads to another classic paradox: the "grandfather paradox." The idea is that a time traveler can go back in time and kill her grandfather before her mother, or she, was born, thus the paradox. Thorne thinks this is too paternalistic and invites the time traveler to kill her mother, giving rise to the "matricide paradox." Novikov argues for leaving out the middleman. Kill your younger self in a time-contorted suicide. The result is the same. The time traveler could not have existed in the first place to commit any of the ansatz-testing crimes.

All these examples invoke people and death to make them graphic, but people raise the issue of consciousness and free will and those issues are messy for a physicist. Joe Polchinski, then of the University of Texas, now at the University of California at Santa Barbara, invented a simple mechanical paradox. Physicists often refer to "pool-ball" physics, meaning the process of reducing a problem to something as visceral as pool balls bouncing off one another so that the physics – conservation of momentum, for instance – can be easily visualized. Polchinski adopted this metaphor to present the "pool-ball crisis." In this thought experiment, a pool ball rolls into one end of a time machine. It comes out the other end in the past. It smacks its earlier incarnation, deflecting it so that it does not enter the worm hole. The paradox is the same in principle. How does the pool ball "get there" in the future if it never entered in the past? Polchinski argued that this simple set up showed that time machines could not exist and no kindly grandfathers or warm, loving mothers were threatened in the least.

The time-machine explorers did not buy it. The flaw in this argument, according to Novikov, is that the original pool ball is pictured as rolling unimpeded into the worm hole, and the collision is only considered when the ball emerges to collide with itself. That is not self-inconsistent. The original pool ball must be involved in the collision as it first rolls toward the opening of the worm hole. Physics must be self-consistent, Novikov insists, even in the presence of time travel. Novikov and his colleagues have carefully studied the pool-ball crisis and have shown that it cannot arise. They have looked at every conceivable interaction. Pool balls can miss, or they can strike a glancing blow, but they can never undergo a hard collision that leads to a paradox. Novikov's group even explored an exploding pool ball, one fragment of which manages to enter the worm hole, come back in time, and hit the exploding pool ball, causing it to blow up, rendering the whole experiment self-consistent. The notion that physics can incorporate time machines in this way is called, in some circles, the Novikov consistency conjecture.

Now we can reintroduce people. According to the consistency conjecture, any complex interpersonal interactions must work themselves out self-consistently so that there is no paradox. That is the resolution. This means, if taken literally, that if time machines exist, there can be no free will. You cannot will yourself to kill your younger self if you travel back in time. You can coexist, take yourself out for a beer, celebrate your birthday together, but somehow circumstances will dictate that you cannot behave in a way that will lead to a paradox in time. Novikov supports this point of view with another argument: physics already restricts your free will every day. You may will yourself to fly or to walk through a concrete wall, but gravity and condensed-matter physics dictate that you cannot. Why, Novikov asks, is the consistency restriction placed on a time traveler any different?

What about the converse? If personal free will exists, does that mean time machines cannot? That question is unresolved. Physics cannot treat the issue of free will, but it may yet address the question of whether time machines can truly exist. The consistency conjecture does say that certain time-travel plots are allowed and others are not. In particular, the consistency conjecture would say that one cannot use time travel to change the future, the basic premise behind both the *Back to the Future* and the *Terminator* movies. Loops in time are allowed, but according to the consistency conjecture, the future is as fixed as the past and cannot be affected by an act of will or any other physical act.

Another way to resolve these issues is to say time somehow "forks off" at the moment of a paradox. The "many world" idea arose in another context as a way to understand some of the conundrums of the quantum theory, how a wave of probability can be turned into an experimentally measured certainty. In the context of time travel, the idea is that in one time prong a time traveler lives on, even having killed her younger self. In this view, her younger self lives in the old time prong, but not in the current one. It is not clear that this resolves the origin of the memories of the time traveler of having been younger and having later wielded the knife.

Philosophical questions aside, the issues involved in time-machine research are right at the frontier of modern physics. We have known since the advent of quantum mechanics that the vacuum does not have zero energy. Having a specific energy, even zero, would violate the Heisenberg uncertainty principle. Rather, the vacuum is riven with fluctuations, particles of light, matter, and antimatter that constantly form and annihilate. The worm-hole mouths, like the space near the event horizon of a black hole, will be endowed with these vacuum fluctuations. In the case of a black hole, these fluctuations lead to Hawking radiation and to the evaporation of the black hole. For a worm hole, the issue is, if anything, even deeper. The vacuum fluctuations can travel in normal space to the opposite mouth of the worm hole, zip inside, and emerge in the past just at the time they left. If that were to happen, there would be twice as much energy in vacuum fluctuations. The cycle might repeat indefinitely and build up an infinite energy density, thus sealing off the worm hole or preventing it from having existed in the first place.

To properly address this issue, a full theory of quantum gravity is required. This theory must incorporate both violently curved space-time and the probabilistic nature of the quantum theory. Such a theory is the holy grail of modern physics. This theory is needed to understand the singularity of the big bang and that inside a black hole. There are great conceptual problems facing the development of such a theory of space-time that applies on scales where time and space themselves are uncertain in a quantum manner, where up and down and before and after lose their meaning. Only with the development of this ultimate theory of everything will we really know whether time machines are conceptually possible.

3. Quantum Gravity

The search for quantum gravity, a theory that unites both the aspects of uncertainty from the quantum theory and the aspects of curved space from general relativity, a theory of everything, is the current frontier of physics. Black holes are at the center of the action. The current contender for this intellectual prize is what is called by physicists *string theory*. The basic notion is that the fundamental entities of the Universe are not particles, dots of matter, but strings of matter, entities with one-dimensional extent.

That seems like a simple, maybe even unnecessary, generalization of our standard picture of elementary particles, electrons, neutrinos, protons, neutrons, and quarks. The doors that have been opened by this change in viewpoint are, however, wondrous.

For perspective, let us go back to the theory of Newton. Newton gave a rigorous mathematical framework in which to understand gravity and much else of basic physics, how things move under the imposition of forces. Newton's law of gravity was based on the concept of a force between two objects. It was encapsulated in a simple formula that said that the force of gravity was proportional to the mass of two gravitating objects and inversely proportional to the square of the distance between them. This prescription was immensely successful. It is still used with great effect in most of astronomy to predict the motions of stellar objects from asteroids to the swirling of majestic galaxies. It is used to guide man-made satellites and rockets. We know now, however, that Newton's theory is wrong. It is wrong in concept and wrong in application.

A hint of the conceptual problem with Newton's theory comes by examining the law of gravity (see also Chapter 11, Section 1). Newton's version of this law tells of the dependence on the masses of the gravitating objects and the distance between them but is mute on the dependence on time. To see this, consider the following question. Take two objects that are attracting one another with their mutual force of gravity. Then, suddenly, move one farther away. How soon does the other object know that the gravity pulling on it is weaker? Newton's formulation demands the answer: no time elapses. The effect is instantaneous. Newton knew that the speed of light was a speed limit, yet his theory demanded communication of information, the strength of gravity, at infinite speed. Another clue to problems with Newton's theory is that if you reduce the distance between two objects to zero, the gravitational force between them is infinite. The lesson of modern physics has been that if your theory gives you an infinite result, in this case infinite speed or infinite force, there is an error in your physics. If one looks sufficiently closely at Newton, those errors exist. The ultimate test is comparison of theory with observation and experiment. Newton is exceedingly successful in many applications but fails in some. The famous case of the shift of the orientation of the orbit of Mercury is one. The angle of deflection of light by a gravitating object is another. Newton's theory gives the wrong answer to these subtle and carefully posed experimental situations.

Einstein's theory of gravity, general relativity, was based on an incredibly simple and elegant idea: that physics should behave the same independent of the motion of the experimenter. The earlier version of this idea, Einstein's special theory of relativity, arose from the young Einstein asking another simple question: what would an electromagnetic wave look like if an observer moved along with it at the speed of light? To answer that question, to show that the observer could not move at the speed of light, Einstein had to show that the speed of light was the same independent of the motion of the observer. This result, one of the deeply true aspects of physics, remains one of the most incredible of human insights. Einstein also proved with his special theory that the lengths and times measured by an observer depended on how the measured object was moving, not in an absolute sense, but moving with respect to the observer.

Einstein's general theory took another step and asked about observers not in uniform motion, the subject of special relativity, but observers in accelerated motion. He realized that an observer freely falling in a gravitational field would measure physical effects and find them identical to an observer moving at uniform speed far from any gravitating object, but that an observer in an accelerating frame would feel exactly the same as one feeling the effects of gravity. This notion has been enshrined as Einstein's *equivalence principle*, that an acceleration gives the same effects as being at rest in a gravitational field. If you sat in a chair in a lecture hall that accelerated at a uniform rate, the floor would push on your feet and the seat would push on your rear end, exactly the same forces you feel sitting in your chair reading this book. The equivalence principle is elegantly simple to state. To put it into a self-consistent mathematical framework, Einstein found that he had to introduce the notions of curved space and a complex set of tensor equations to describe it. Our sense of the nature of space has never been the same.

Einstein's theory of gravity has passed every test put to it. It gets the right answers for the shift of Mercury's orbit and the deflection of light, and has passed numerous other tests to the limit of our current ability to devise those tests. This makes general relativity a better theory of gravity than Newton's. General relativity also becomes identical to Newton's theory, mathematically, and hence in its precise predictions, when gravity is weak, distances are large, and motion is small. It must do so in order to reproduce Newton's manifest success of predictability in those regimes. To accomplish this great success, Einstein had to abandon, not just the mathematical structure adopted by Newton, but the fundamental concept behind gravity. Einstein abandoned the notion of a "force" of gravity, and replaced it with the notion of curved space and warped time. Space is curved, and that tells matter how to move, to orbit, to fall. Gravity is geometry, the geometry of curved space. The change in conception wrought by Einstein was deeply profound. General relativity is, however, wrong.

So far we only know that general relativity is wrong because of conceptual problems. We have not been able to devise a test sensitive enough to display the fact. The conceptual problem is in the prediction of the singularity. General relativity predicts that, right at the center of a black hole, a region of infinitesimal size, with infinite space-time curvature and infinite tidal forces must form. Those predictions of infinity are the undoing of general relativity. As a predictive theory, general relativity is fine in the regimes where it works, just as Newton's theory was in its own regime. General relativity does everything that Newton's theory could do and more, including predictions of black holes and event horizons.

A theory of everything must take its place in this hierarchy. It must incorporate everything that Newton accurately predicted. It must also incorporate everything that Einstein consumed so elegantly. Then it must also answer the question: what is this amazing thing called a singularity? The theory must tell us what happens to space and time under conditions where quantum uncertainty dictates that the very notions of "front," "back," "here," "there," "before," and "after" lose their meaning. There must be space without space as we know it and time without time as we know it. Is there any wonder that physicists since Einstein have labored against immense conceptual problems in attempting to cross this barrier?

4. String Theory

Work on string theories is beginning to penetrate these barriers. The previous summary of the history of this development gives some preparation for what is necessary. Whereas Einstein overthrew the concept of gravity as a force between two objects, the theory of everything is likely to bring with it entirely new ways to think about gravity and, indeed, about space and time. In the appropriate regime, one can still think of curved space as the origin of gravity, just as for weak gravity it is still useful to think of a force of gravity and to use Newton's theory in appropriate circumstances. One of the steps that energized string theory was the understanding that within the full mathematics of the theory, a subset described exactly Einstein's theory of general relativity. Just as Einstein's theory "contains" Newton's theory of gravity in the limit of weak gravity, string theory contains Einstein's theory. String theory, however, holds a lot more. The underlying concepts of a theory of everything may require a shift in conceptual basis as profound as that from a force of gravity to gravity as curved space. Recent developments point strongly to the conclusion that, at a sufficiently small scale, where neither Newton nor Einstein dared tread, neither space nor time exist.

To see how this idea has arisen, a sketch of string theory is necessary. The basic notion is that particles, mathematical points, are too simple to contain the wonders of nature. True point particles have no inner structure, no richness. A string, on the other hand, by adding only one more dimension to the structure, can vibrate in many modes. You can't make music with four grains of sand, but with four violin strings you can have Mozart! In the view of string theory, different modes of vibrations of the string represent different particles, just as one string on a violin can give different notes depending on where the violinist's finger is placed.

Unlike violin strings, the strings that represent the fundamental entities in this theory do not exist only in our ordinary three-dimensional space. To make a mathematically self-consistent picture, one free of infinities and other inconsistencies, the space through which the strings thread must be of much higher dimension. The currently most viable versions of the theory have ten or eleven dimensions.

To illustrate black holes and curved space, we have had recourse to embedding diagrams that reduce the fullness of the curved three-dimensional space to two so that we, as three-dimensional creatures, can view these warped spaces from our higher-dimensional perspective (Chapter 9, Section 5; Chapter 11, Section 1.3; Figure 12.2). From this perspective, it is clear to us that, even though there is no two-dimensional outside to the two-dimensional space, there is a very natural "outside" to the two-dimensional space, the very three-dimensional "hyperspace" that we occupy. This naturally leads one to wonder whether there is a "real" fourth spatial dimension that we, as three-dimensional creatures, cannot perceive, into which our three-dimensional Universe curves. This hyperspace would be where worm holes go when they go.

Despite the intuitively natural sense that invokes this sort of higher dimension, this is not what the string theorists are talking about. Physicists can construct mathematical models of curved three-dimensional spaces and universes, even worm holes, completely within the confines of that three-dimensional space. There is no need, or means, to invoke any higher dimension, no way to measure it, no way to do physics with it. Not yet, anyway.

Rather, the higher dimensions invoked by string theorists are all "compact." To picture a compact space, start again with a two-dimensional analog, a sheet of paper. As shown in Figure 12.3, roll the paper up into a tight roll. From a distance, the resulting object looks like a straight line, a string of length of perceptible extent, but no width. Imagine rolling the paper up laterally so you have a tiny ball. Now from a distance, the whole original sheet of paper resembles a point, a particle of no extent. A string in that original sheet of paper could still exist and vibrate away in that compact space that we could not directly perceive. We could, however, deduce that the higher dimensions exist because the nature of particles in our Universe demands it!

The last few years have seen some immense advances in string theory that have given great hope that it is the basis for the theory of everything. One step has been to prove that what looked like five or six different string theories are all versions of the same underlying theory, the full shape of which has not yet been elucidated. These connections were established by what physicists have called *duality*, a connection between the properties of the theories. In one version of the theory, a parameter could be small, and, as the parameter got large, the mathematics of the theory broke down. In another string theory, the dual to the first, there would be a parameter that was just the inverse of the first. In that second theory, as the parameter got large, the inverse parameter got small, and the mathematics in that theory was well behaved. The middle ground is unknown, but this duality yields a signpost for how to link the disparate theories and show that they are deeply connected, that they are aspects of the same thing. This grand string theory that is taking shape is called *M theory*, *M* for matrix.

One of the concepts that has emerged from string theory is that there are not only strings threading the ten or eleven dimensions of the string theory hyperspace but also surfaces. These surfaces can be canted in hyperspace in just the same way that a sheet of paper can be oriented in all sorts of ways in our ordinary three-dimensional space. A more general word for a surface is a membrane, a term that also connotes a certain elasticity, a property that these surfaces have. These membranes can vibrate just as the strings can vibrate, and their

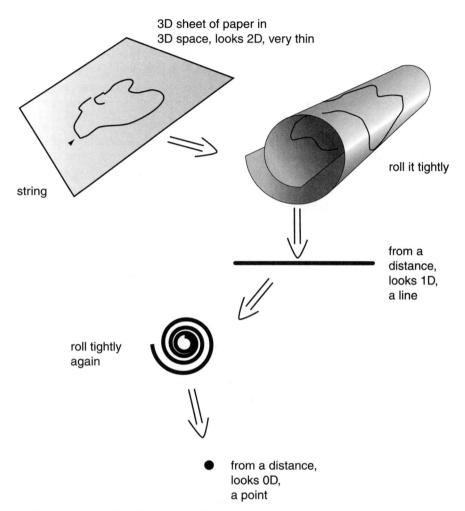

Figure 12.3. A schematic example of how a space could be compact and still contain a string capable of vibrating. A two dimensional sheet containing a one-dimensional string can, in principle, be rolled up compactly so that it would appear to have only one dimension, length. The one-dimensional string would still be there, just wound up in the compact space. If the space were rolled up again, it could, in principle, be compacted into a point, a one-dimensional space, yet still contain the string.

modes of motion are also important to the behavior that emerges as ordinary physics in our ordinary space time. To classify the membranes in spaces of various dimensions, they are referred to as *p*-branes, where *p* is a symbol denoting the dimension of the membrane, *p* = 2 for a two-dimensional surface, *p* = 3 for three-dimensions, *p*

= 10 for ten dimensions. An important development of string theory in recent years has been the recognition of the critical nature of the interaction of strings with *p*-branes. The ends of the string can attach to the *p*-branes or snap off to form closed rings. Much of this work was done by Joe Polchinski at the University of California at Santa Barbara, and his colleague, Gary Horowitz, but many others are contributing to the fevered pace of development.

A striking feat that followed the development of the theory of *p*-branes and their interactions with strings has been the capacity to construct simple models of black holes. These black holes are not the creatures of the curved space-time of Einstein, but simpler versions in two dimensions constructed from the entities of *p*-branes and strings. Nevertheless, because string theory contains Einstein's theory, objects that exert gravitational pull and that have event horizons can be constructed. The difference is that string theorists can count the numbers of modes and vibrations of the strings within the black holes they have constructed and tell exactly what the temperature and entropy should be. They get precisely the same answer as Hawking did in predicting Hawking radiation (Chapter 9), even though the mathematics and, indeed, the conceptual framework they use, is completely different. This striking concordance is the sort of development that tells physicists that they are getting close to a universal truth and that string theory has deep lessons to reveal.

String theory has also brought new insight into another problem that arises from thinking about the nature of black holes. This is called the *information crisis*. Information, the bits and bytes of computers, is about as fundamental as you can get. The problem is that black holes seem to destroy information, and that bugs physicists. The idea was already there in our previous discussions of the nature of black holes in Chapter 9 and captured in John Wheeler's phrase "black holes have no hair." You can throw stars, cars, people, and protons into a black hole, and all the information that described that ordinary stuff vanishes inside the event horizon. The only properties of a black hole that can be measured from the outside are its mass, spin, and electrical charge. Now Stephen Hawking enters the game. Black holes can evaporate, giving off Hawking radiation. Given enough time, the black hole will just disappear, leaving pure radiation with very little information content, essentially pure randomness. This process conserves energy, the energy equivalent of all the stuff that went down the black hole eventually emerges as the energy in the radiation. What happened to the information that defined that stars, the cars, the people, and the protons that went down the hole? Physicists have been debating this fundamental problem since the implications of Hawking's ideas of black hole evaporation were first assimilated.

One can sense a possible wrinkle in this argument. Hawking's theory was designed to work for ordinary-size black holes where the event horizon was well separated from the singularity at the center of the black hole. When a black hole evaporates down to the last of its essence, one needs a theory that can simultaneously treat the event horizon and the singularity and that probably requires a quantum gravity, a theory of everything. In the absence of that theory, it is not clear that one can use Hawking's original theory to account for the final moments. String theory gives a different possibility. It suggests that the black hole cannot evaporate entirely, but that, as the process runs away, one is left with a string vibrating intensely somewhere in its ten-dimensional space-time. In those vibrations is the epitaph of all that entered the black hole, all that original information, the size of the stars, the bumper stickers on the cars, the personalities of the people, the number of protons.

Einstein wrote down a full and self-consistent set of equations to describe gravity (in the absence of quantum effects) in 1916. Those equations have yet to be fully solved. String theory is like that, only more so. The full mathematical structure of string theory is very complex, and only a few solutions have been wrested from it. Those solutions have been tremendously encouraging. Exactly what theory of space and time will emerge from string theory is thus not yet clear. One can see that, because string theory is a theory of quantum fields and forces, the fundamental concept of gravity will again be a force, but a quantum force, not that of Newton. Away from any singularity, this "force" of gravity will act just as in Einstein's theory. One will be able to speak in the language of curved space and time and dream of the construction of worm-hole time machines.

On the microscopic scale, however, the new concepts of string theory will lead to different pictures, pictures that are only just now beginning to take hazy conceptual form. One can see that gravity will be represented by the familiar terms of Einstein's gravity plus "something else" that comes in ever more strongly as one approaches, intellectually at least, the singularity. At the singularity itself, Einstein's theory will be completely inapplicable, as Newton's theory is within the event horizon of a black hole. At the singularity, space and time as we think of them on our large, human scale will probably not exist.

Even with these encouraging clues and the startling successes of string theory, one can see that we have a long way to go in our intellectual quest to understand the Universe.

5. When the Singularity Is Not a Singularity

The singularity of Einstein's theory cannot exist. Something else must happen to space and time "there." In the absence of the full development of a string theory or some other theory of everything, physicists are left to grope. When physicists grope, startling ideas emerge.

We know the scale on which Einstein's theory must break down even if we do not fully understand what must replace it. This scale can be estimated from the simple idea of asking about the conditions where quantum uncertainty must be as important as the space-time curvature of gravity. The fundamental constants of quantum gravity are the strength of gravity as measured by Newton's constant from the world of the large, the degree of quantum uncertainty as measured by Planck's constant from the world of the small, and nature's speed limit, the speed of light from the world of the very fast. With values for these constants of nature in some set of units, English or metric, it does not matter, one can estimate the scale where Einstein's theory, and ordinary quantum theory, fail. This scale, of length, time, density, is called the *Planck scale*. Newton's constant has units of length cubed, time squared, and the inverse of mass. Planck's constant has units of mass, length squared, and the inverse of time. The speed of light has units of length over time. There is only one way we can combine these three fundamental constants with their individual units to produce a quantity of only length, only one other way to produce a time, and only one third way to produce a mass. This exercise is a simple one of sorting out units, but it has profound implications because the building blocks are the fundamental constants that tell us how space curves, the degree of quantum uncertainty, and how fast things can move. Their combination implicitly tells us where space gets so curved that a quantum wave cannot exist and simultaneously where quantum uncertainty is so large that speaking of a given curvature makes no sense. We learn the conditions where the two great theories of twentieth-century physics butt heads and contradict one another, the conditions that call for a new theory of physics.

The resulting value of the length, the Planck length, is about 10^{-33} centimeters. This is an incredibly small value, much smaller than the size of a proton, but it is not zero! This is roughly how large the singularity must be. At this level, space and time break down into something else, and Einstein's prediction of a singularity goes awry. The corresponding Planck time is about 10^{-43} seconds. This is again an incredibly short time, but not zero. Time as we know it probably does not exist at shorter intervals so that asking what happened when the Universe was younger than 10^{-43} seconds or before the big bang may not make sense, at least not in the traditional way. The Planck mass is about 10^{-5} grams. This is a small number, but not incredibly small. It is vastly bigger than any elementary particle we know. One can also work out the Planck density, the Planck mass divided by the cube of the Planck length. The answer is about 10^{93} grams per cubic centimeter. This is a gigantic density, but it is not infinite. In some average way, this must be the density of a singularity, the density from which our Universe expanded in the big bang, the density to which all is compressed in the centers of black holes.

One way to think about the singularity is as a bubbling sea of Planck masses, each a Planck length in extent winking in and out of existence for intervals of a Planck time. This quantum-bubbling mess has been called a *quantum foam*, another bit of etymological brilliance from John A. Wheeler. This term is a picturesque name intended to describe something we do not understand, yet capture the flavor of the idea that it is not ordinary space and time. In the quantum foam, one could not speak of front and back because space itself would be so quantum uncertain that such concepts are invalid. The same is true for the ideas of before and after, with time a quantum froth.

Even in the absence of a full theory, if we picture the singularity not as a point of zero size and infinite density but a dollop of quantum foam, then other ideas begin to emerge. The Universe was not born from a point of infinite density but emerged as a bubble of ordinary space and time from this quantum foam. This bubble was highly energetic and expanded to become everything we see. Again, as we discussed in Chapter 11, the expansion is pictured in the sense that all points of space move away from all other points of space, not an explosion of stuff into a preexisting space. Also, as three-dimensional physicists, we do not have to address the issue of what the three-dimensional Universe is expanding into, as much as that question seems to intrude.

That the Universe emerges from the quantum foam already gives some predictability to the nature of the Universe. There must have been quantum fluctuations in the density and temperature of the very young, hot big bang as it emerged from the quantum foam 10^{-33} centimeters across and 10^{-43} seconds old. These unavoidable fluctuations can be calculated from the quantum theory with some assumptions, and they later cause the tiny irregularities in temperature that *COBE* detected (Chapter 11) and that later grow to form all the structure we see – stars, galaxies, clusters of galaxies.

The notion of a quantum foam also plays a role in the thinking about worm holes and shows again that we cannot pursue the physics of worm holes without a theory of the quantum foam, a theory of the singularity, a quantum-gravity theory of everything. One way to picture the quantum foam is as quantum-connected fragments of space and time, connecting different places and different times willy-nilly in a probabilistic way. These connections, although dominated by quantum uncertainty, are essentially tiny quantum worm holes. One can imagine making a worm hole by taking a little quantum loop of space and time and blowing it up to become a worm hole big enough to travel through.

Another way to imagine making a worm hole leads to similar issues of the quantum nature of space and time. If you start from ordinary space and want to make a black hole, you have to stretch and distort the space, but you do not have to rip or tear it (at least not until you get to that nasty singularity). That is not true for a worm hole. To make a worm hole, you have to tear and reconnect space. You have to change not just the curvature of space but its connectedness, its topology. If you think about it, a tea cup with a nice handle and a donut are the same basic thing in terms of how they are connected. They are both solid objects with one hole through them. You could make both from the same lump of clay by just molding a side of the donut shape to be the cup and shape the clay around the hole to be the handle. You would not have to tear the clay or reattach it at any point. You cannot, however, make a solid lump of clay into either a donut or tea cup without tearing a hole in the clay.

Think of how you could connect space on a large scale to make a worm hole. It helps to imagine this in two dimensions. Picture a balloon. Push two fingers inward from opposite sides until your fingers almost touch, separated only by the thin rubber of the balloon. You have almost made a worm hole. If the connection could be made there in the center of the balloon, there would be a way to travel on a shortcut through the center of the balloon, rather than taking the long way around on the surface. The balloon serves as a two-dimensional analog of our three-dimensional space, so all motion is confined to the rubber of the surface. Now think of what you need to do to make the connection between your fingers. You would have to cut the rubber and attach the ends of the two cones; but cutting the rubber is the analogy of cutting the very fabric of space. That would be the issue in our real three-dimensional space in order to make a three-dimensional worm hole. The cutting and reattaching of space would amount to, at least temporarily, introducing an end to space, a singularity, before the reattachment is made. To make a worm hole or a worm-hole time machine in this way, we have to bring in the operation of introducing a tear in space-time, a tear in the quantum foam. We will not know whether such an operation even makes sense until we have a theory of quantum gravity that tells how space and time behave if such a rent is threatened. Once again, we cannot think constructively about worm holes or time machines without a theory of quantum gravity to guide us.

If the Universe were born not from a singularity of infinite density, but from a spot of quantum foam, then the inverse is true. When a star collapses to make a black hole, the matter of the star does not disappear into a singularity of zero volume but is crushed into a froth of quantum foam of a Planck density. One of the most dramatic ideas to emerge in the last few years was to ask, if a black hole leads back to the quantum foam from which the Universe arose, why cannot the cycle repeat. This idea was first put forth by Andre Linde, a Russian physicist, now at Stanford University. Linde was striving for some new idea to present at a conference to which he had been invited. He was ill and contemplating skipping the meeting, when this notion came to him. He worked out the basic mathematical and physical picture and presented it at the meeting.

The idea is that the quantum foam that forms at the center of the black hole is identical to that from which the big bang, our whole Universe, arose. This means, Linde argued, that a new Universe can arise from the quantum foam of the black hole. The dramatic implication is that the chain could be endless. A universe forms; it expands to form stars. Some of the stars collapse to make black holes. From the singularities of those black holes, new Universes can be born. Here, perhaps, is a way to answer the question of what came before and what comes after the big bang – endless Universes forming endless black holes.

Like many grand ideas of physics, this one must be poked and pummeled and analyzed. How do you prove such a startling conjecture? We cannot travel to other Universes to see how they work. We are stuck in this one but empowered with our imaginations and our mathematics and physics. Physicists are already at work generalizing the old cosmologies to see how these ideas could fit in. The easiest way to picture a bubble being blown in the quantum foam to become our Universe is to picture a literal bubble being blown. Such a bubble, basically a sphere, is a two-dimensional analog, an embedding diagram, for a closed three-dimensional Universe. Such a Universe would have a finite lifetime and would have to recollapse. That is in conflict with the favored model of an "inflating" Universe, which only makes conceptual sense if the Universe were flat, the three-dimensional equivalent of an infinite, flat plane. The results reported in Chapter 11 suggest that the Universe is not closed and "spherical." It might be flat, but accelerating or open and doomed to expand forever, a saddle shape in the two-dimensional analog. Physicists and cosmologists are working now to develop models of inflating Universes that are consistent with infinite expansion. Such universes, can, of course, make black holes as they expand, and that is enough to raise Linde's conjecture of new universes being constantly created.

These ideas have been taken one more dramatic step by Lee Smolin of Penn State University in his book, *The Life of the Cosmos*. Smolin addresses the deepest issue that drives both physicists and theologians. Why are *we* here? What is it about our Universe that gave rise to life, to us. Smolin may not have the answer, but he has put the issues in an especially thought-provoking way by combining these ideas from physics with the basic ideas of biology, the power of natural selection. Smolin notes the amazing coincidences of numbers and physical conditions that are required to give rise to life as we know it. What if, Smolin wonders, each new universe had different numbers, for instance different values of the fundamental constants, Newton's constant of gravity, Planck's constant, the speed of light, and other physical constants of nature. Most of those universes would fail. Some would not get out of the quantum foam or would quickly fall back.

Others would expand so rapidly that stars did not have a chance to form, so there would be no black holes. In either case, those universes would be barren, unable to produce progeny, new universes with new properties. Smolin makes a natural selection argument that after countless trials, the universes that survive would be those that maximize the production of black holes so that maximum progeny are ensured. Smolin argues that physicists may have to give up on a purely reductionist approach to science wherein the constants of nature have set values that theory and experiment can reveal and accept that our Universe has arisen from a process of trial and error, a result of probabilities, not certainty. To be fruitful, such a universe would have to expand about as fast as ours, make stars like ours, produce heavy elements like ours to control the heating and cooling of the interstellar gas to keep star formation going for billions of years. Such a universe, Smolin deduces, must have the properties of our Universe, and such a universe naturally gives rise to life to contemplate and make sense of it. Now that is a grand vision.

For all its inventiveness, Smolin's picture does not really address the fundamental issue. Given that there are infinite universes experimenting with all possible forms, how did it all arise in the first place? Was there a beginning to this process? Is there an end? James Gott of Princeton has put another wrinkle on the game by combining the self-reproduction of universes through black holes with the notions of time machines. If new universes emerge from the quantum foam of a black hole singularity, can they emerge in the past? If that were possible, Gott conjectures, then the universe that emerges from a black hole could be the one that made the black hole from which it emerged, or a universe somewhere back in the chain of universes that Linde and Smolin contemplate. Recall that the Novikov consistency conjecture does not rule out time travel, it only demands self-consistency. Could it be that the Universe or a complex web of universes gave rise to itself in a closed but self-consistent time loop? Could it be that there is no "beginning" and no "end" but just an infinite closed loop? As Gott asks, could the Universe have created itself?

This is heady stuff. It is amazing that these ideas have emerged, not from science fiction, but from hard-nosed physicists wrestling to make sense of the Universe of our observations. Progress can be made by examining these ideas for self-consistency, and that enterprise will go forward with great energy. The real solution, or at least the one we can contemplate today, is to develop the theory of quantum gravity, the theory of everything. Today the best bet for that appears to be string theory, M theory. So one can ask, what does string theory say about the quantum foam? Quantum foam was just a name, a place holder until some physics came along. What exactly does string theory say about the conditions at the Planck scale? Does string theory allow new universes to be born from the conditions predicted by string theory for "not time" and "not space" at the center of a black hole constructed from strings?

Other, more speculative questions also arise. What are these higher dimensions that are forced on the string theorists by mathematical self-consistency? Do they simply dictate the properties of particles that appear in the three-dimensional Universe of our space-time, or can they be manipulated in some way? Does string theory allow worm holes and time machines? Does it prevent them?

The first papers discussing the impact of these ideas on the "real world" are starting to appear. One argues that, when a neutron star forms in a supernova, some of the energy of gravitational collapse may be siphoned off into a few of the higher dimensions of string theory. This may be fantasy, but I am sure this contribution to the literature is a harbinger of things to come. It is somewhat old fashioned, but my guess is that even with a theory of everything in sight we are not about to see the end of physics.

Index